DIGITAL_HUMANITIES

ANNE BURDICK JOHANNA DRUCKER PETER LUNENFELD
TODD PRESNER JEFFREY SCHNAPP

THE MIT PRESS CAMBRIDGE, MASSACHUSETTS LONDON, ENGLAND

MIT Press books may be purchased at special quantity discounts for business or sales promotional use. For information, please email special_sales@mitpress.mit.edu or write to Special Sales Department, The MIT Press, 55 Hayward Street, Cambridge, MA 02142.

This book was typeset in Bembo, Benton Gothic, and Knockout by the authors and was printed and bound in the United States of America.

Library of Congress Cataloging-in-Publication Data

Digital_humanities / Anne Burdick ... [et al.].
p. cm.
ISBN 978-0-262-01847-0 (hardcover : alk. paper)

1. Humanities—Electronic information resources.
2. Humanities—Computer network resources.
I. Burdick, Anne.
AZ195.D54 2012
001.30285—dc23 2012026514

10 9 8 7 6 5 4 3 2 1

PREFACE

WE LIVE in one of those rare moments of opportunity for the humanities, not unlike other great eras of cultural-historical transformation such as the shift from the scroll to the codex, the invention of moveable type, the encounter with the New World, and the Industrial Revolution. Ours is an era in which the humanities have the potential to play a vastly expanded creative role in public life. The present volume puts itself forward in support of a Digital Humanities that asks what it means to be a human being in the networked information age and to participate in fluid communities of practice, asking and answering research questions that cannot be reduced to a single genre, medium, discipline, or institution. Digital Humanities represents a major expansion of the purview of the humanities, precisely because it brings the values, representational and interpretive practices, meaning-making strategies, complexities, and ambiguities of being human into every realm of experience and knowledge of the world. It is a global, trans-historical, and transmedia approach to knowledge and meaning-making.

Yet there remains a chorus of contemporary voices bewailing yet another "definitive" crisis in humanistic culture, yet another sacrifice of quality on the altar of "mere" quantity. Our response is not just a counterargument in favor of new convergences between quality and quantity, but also one in favor of a model of culture embodied by this book itself. We do not think the humanities are in perpetual crisis or imperiled by another battle for legitimacy with the sciences. Instead, we see this moment as marking a fundamental shift in the perception of the core creative activities of being human, in which the values and knowledge of the humanities are seen as crucial for shaping every domain of culture and society.

The model we have created is experimental. It moves design—information design, graphics, typography, formal and rhetorical patterning—to the center of the research questions that it poses. It understands digital and physical making as inextricably and productively intertwined. This model is collaborative and committed to public knowledge. Crafted for a heterogeneous audience with crisscrossing and even contradictory interests and needs, it is meant as a porous multiple construct: a guidebook for the perplexed, a report on the state of the field, a vision statement regarding the future, an encouragement to engage, and a tool for critically positioning new forms of scholarship with respect to contemporary society.

What you are reading is a collaboratively crafted work. Each of the authors contributed to the writing, editing, reworking, and final composition of the whole. Each has brought to this endeavor theoretical and conceptual engagements based upon personal experience, a commitment to experimental forms of scholarship, and expertise working with media, developing digital platforms, and engaging in design practice.

The first three chapters offer synthetic mappings of the field, its emerging methodologies, and its social characteristics. Chapter 1. **HUMANITIES TO DIGITAL HUMANITIES** explores emerging forms of transmedia research and the increasing importance of prototyping, experimentation, and tool and platform development for contemporary scholarship in the humanities. Chapter 2. **EMERGING METHODS AND GENRES** charts new ways of doing things using digital tools and platforms that extend traditional scholarly practices or devise entirely new ones (whether new fields of inquiry or new models of dissemination and practice): the shapes that scholarly knowledge can assume in digital environments, the models of practice that are becoming prevalent, and the units of argument of which they are composed. Chapter 3. **THE SOCIAL LIFE OF THE DIGITAL HUMANITIES** analyzes the real and potential roles that Digital Humanities projects are playing in contemporary society, the purposes they serve, the communities engaged by them, and the values they affirm.

Chapter 4. **PROVOCATIONS** builds from the synoptic to offer a series of propositions regarding what the future might hold for the Digital Humanities specifically and the humanities generally. The conclusions are speculative, raising thorny questions whose answers necessarily lie beyond the scope of present knowledge.

In addition to the main chapters, there are two other components to this book. At the end of Chapter 2, we offer **A PORTFOLIO OF CASE STUDIES** for launching Digital Humanities projects into the world. To avoid forming an inadvertent canon, we have chosen *not* to pick from the host of exemplary Digital Humanities projects that already exist, many of long-standing impact and significance. Instead, we chose to aggregate and synthesize their defining features in the form of speculative case studies, fictions that delineate emerging methods and genres informed by present capabilities and resonant with the lessons gleaned from several decades of pioneering work. Our goal here and elsewhere in this book is to provide a concrete framework for the creation of generative scholarship. The case studies provide models for building teams, assembling necessary technical resources, and situating the projects within cross-disciplinary and multi-institutional configurations.

Following Chapter 4, we present **A SHORT GUIDE TO THE DIGITAL_HUMANITIES**. Here we condense the arguments in the book into three sets of **QUESTIONS & ANSWERS** that serve both the initiated and the novice alike. We provide a short overview of the fundamentals of the Digital Humanities, answer why projects are the basic unit for Digital Humanities scholarship, and describe the institutional relationships that grow out of and contribute to Digital Humanities work. Understanding how and according to what criteria these modes can be evaluated have become matters of institutional urgency. To this end, **A SHORT GUIDE** identifies five areas in which Digital

Humanities work is currently being produced and assessed, translating each of them into a checklist of items referred to as **SPECIFICATIONS**. The specifications outline the ethics, values, methods, and best practices for student and faculty researchers, staff, administrative officials, faculty committees, and others involved in the production, oversight, or review of digital projects. They are intended as guides for parties interested in designing and fostering project-based scholarship, determining core competencies and methods, adopting assessment criteria for digital work, measuring outcomes, and engaging in advocacy for the Digital Humanities. We are openly distributing **A SHORT GUIDE** on the Web and via social media as we hope that this compact form will serve to bring colleagues and students up to speed, and offer support to those who are charged with evaluating emerging Digital Humanities work within existing institutional frameworks.

AFTERWORD: NOTES ON PRODUCTION details how the book's collaborative authoring and design process evolved, and what lessons it might offer to others interested in pursuing new modes of knowledge formation. Finally, **REFERENCE NETWORKS** points outward, linking this book to the discourses and practices of this dynamically changing field.

The attentive reader will already have noticed that *Digital_Humanities* is not a standard-format academic work. It is not a collection of individually authored scholarly papers or research reports on the history of, or critical engagement with, the Digital Humanities. Neither is it a textbook from which to teach the discipline's foundations nor a manual of technical specifications, much less a discussion of every facet of the field, its protagonists, successes and failures, and defining moments. In lieu of a bibliography, it includes a "reference network" and list of works cited.

That is because the Digital Humanities remains at its core a profoundly collaborative enterprise. Over the decades, a diverse array of individuals, projects, and organizations has built the field of Digital Humanities as it exists today. We cite some of these precursors and colleagues in our text, while many more of them inform the book's ideas and arguments. In shaping this volume, we have striven not to privilege one lineage or another within the Digital Humanities, seeking instead an encompassing yet polemical voice that speaks both inside and outside the walls of the academy. Accordingly, our case studies are fictional, our quotations of specific figures and theorists sparing, and our language direct.

This book is a compact work composed by a group of practitioners from a variety of humanities disciplines and fields, including design. For all their diversity, the authors share a core conviction that informs every page of the book. This core conviction is embodied in the title of our book, *Digital_Humanities*. The underscore between the two words references the white space between them as a vital yoke and shifting signifier, one that presents the two concepts in a productive tension, without either becoming absorbed into the other. The underscore is not merely a graphical notation; rather, it is used deliberately as an overdetermined marker of the critical nexus between "digital" and "humanities." It references the precarious, experimental, and undefined future of the humanities in a world fundamentally transformed by everything digital. Although we do not use the underscore throughout the text, it remains the subject of every page of this book. And while it may seem paradoxical to write a book called *Digital_Humanities*, the very act demonstrates the continuities that link current practice to long-standing traditions.

We graciously acknowledge the support of the Andrew W. Mellon Foundation and the UCLA-based Andrew W. Mellon Foundation Grant for Transformational Support in the Humanities led by Ali Behdad and Timothy Stowell as well as participants in the 2008-09 seminar "What is(n't) Digital Humanities?"

Earlier versions of some of the ideas in this book were expressed in the UCLA white paper "The Promise of Digital Humanities," co-authored by Todd Presner and Chris Johanson, as well as the proposals for the UCLA Digital Humanities minor and graduate certificate, co-authored with Johanna Drucker, Diane Favro, Chris Johanson, Todd Presner, Janice Reiff, and Willeke Wendrich. The authors also thank David Shepard and Miriam Posner for their critical feedback.

The specification "How to Evaluate Digital Scholarship" includes contributions and language provided by John Dagenais, Diane Favro, and Willeke Wendrich.

We thank the graduate program in Media Design Practices at Art Center College of Design for providing space and resources, and in particular, for the fresh perspective of graduate research interns Brooklyn Brown, Bora Shin, Matthew Manos, and Jayne Vidheecharoen.

Orli Low's copy-editing not only cleaned up our prose, but her questions sharpened our thoughts. Thanks to Doug Sery, our editor at the MIT Press, who has long championed the intersections of academic inquiry and generative practice.

1. HUMANITIES TO DIGITAL HUMANITIES

DIGITAL HUMANITIES IS BORN OF THE ENCOUNTER BETWEEN TRADITIONAL HUMANITIES AND COMPUTATIONAL METHODS.

WITH THE MIGRATION OF CULTURAL MATERIALS INTO NETWORKED ENVIRONMENTS, QUESTIONS REGARDING THE PRODUCTION, AVAILABILITY, VALIDITY, AND STEWARDSHIP OF THESE MATERIALS PRESENT NEW CHALLENGES AND OPPORTUNITIES FOR HUMANISTS. IN CONTRAST WITH MOST TRADITIONAL FORMS OF SCHOLARSHIP, DIGITAL APPROACHES ARE CONSPICUOUSLY COLLABORATIVE AND GENERATIVE, EVEN AS THEY REMAIN GROUNDED IN THE TRADITIONS OF HUMANISTIC INQUIRY. THIS CHANGES THE CULTURE OF HUMANITIES WORK AS WELL AS THE QUESTIONS THAT CAN BE ASKED OF THE MATERIALS AND OBJECTS THAT COMPRISE THE HUMANISTIC CORPUS.

CONFRONTING the massive transformation of knowledge, society, and culture that is underway in the digital age, this book takes stock of this new world as well as anticipates future developments in the Digital Humanities. Building on earlier generations of computational approaches to humanities research—with emphasis on the creation, preservation, and interpretation of the cultural record—the Digital Humanities has greatly expanded the potential power and reach of the humanities disciplines, both within the academy, and, just as importantly, outside its walls.

Even though we recognize the game-changing implications of the adjective "digital," it is on the "humanities" that our attention is concentrated in this chapter. As they developed from their classical and early modern precursors, the disciplines that make up the modern humanities—including, but not limited to, literature, philosophy, classics, rhetoric, history, and studies of art, music, and design—have sought to define culture and help us gain a greater understanding of the human experience. The humanities are siblings of the sciences in their embrace of intellectual rigor and free inquiry. But while the humanities do not shun empirical methods, they have rarely been characterized by the strictest forms of empiricism. Within their fold there has not only been room but also a sense of urgency regarding the need to confront questions of worth, cultural significance, and deeper meaning. Humanists engage with questions of value and interpretation, with the realms of rhetoric as well as logic, with subjective judgment alongside attention to verifiable truths. The spectrum of humanistic thought, like that of scientific investigation, encompasses the gamut of beliefs regarding the nature of knowledge, the world, and the human ability to establish understanding with various degrees of certainty. Digital capabilities have challenged the humanist to make explicit many of the premises on which those understandings are based in order to make them operative in computational environments.

This chapter opens with a discussion of what precisely we mean by the humanities in the broadest sense and then moves on to a historical account of the earliest attempts to meld humanistic inquiry with digital technologies. In moving past the first generations of Digital Humanities practice, we shift to outlining the implications of design, and specifically design for transmedia modes of argumentation, as a model for contemporary work. The emphasis on design depends on robust technological environments in order to manifest across media, so we discuss how the basics of computation and processing affect the design and implementation of Digital Humanities projects. These projects engage with any number of different methodologies and approaches, but here we concentrate on four: curation, analysis, editing, and modeling as central to contemporary humanistic inquiry. These intertwinings of scholarly method, computational capacity, and new modes of knowledge formation combine to make possible what we term the Generative Humanities, a mode of practice that depends on rapid cycles of prototyping and testing, a willingness to embrace productive failure, and the realization that any "solutions" generated within the Digital Humanities will spawn new "problems"—and that this is all to the good. Finally, we conclude this chapter by making the argument that the Digital Humanities may well function as a core curriculum for the 21ST century.

From Humanism to the Humanities

While the foundations of humanistic inquiry and the liberal arts can be traced back in the West to the medieval trivium and quadrivium, the modern human sciences are rooted in the Renaissance shift from a medieval, church-dominated, theocratic worldview to a human-centered one. The philosophical systems of Renaissance thought, mirrored in the graphical structure of monocular perspective, had human subjects at their core. The gradual transformation of early humanism into the disciplines that make up the humanities today was profoundly shaped by the editorial practices involved in the recovery of the corpus of works from classical antiquity, many preserved in Greek and Arabic manuscripts in Byzantine and Islamic centers of learning. As the first universities were established in the High Middle Ages and monastic scriptoria were joined by university-based scribal shops working under the *pecia* system as well as by courtly scriptoria, a publishing industry arose that fostered a reading public interested in secular as well as scientific and literary matters. The development of vernacular languages and literary forms further expanded the compass of humanistic expression, with the poetry of Dante, Petrarch, and Chaucer as well as the translations of texts from Latin, Greek, Old English, Norse, French, German, and other languages finding their place alongside the classical canon. The wellsprings of humanism were fed by many sources, but the meticulous (and, sometimes, not-so-meticulous) transcription, translation, editing, and annotation of texts were their legacy. The printing press enabled the standardization and dissemination of humanistic cultural corpora while promoting the further development and refinement of editorial techniques. Along with many other scholars, we suggest that the migration of cultural materials into digital media is a process analogous to the flowering of Renaissance and post-Renaissance print culture.

The shift from humanism to the institutionally sanctioned disciplinary practices and protocols that we associate with the humanities today is best described as a gradual process of subdivision and specialization. Carried out in the course of the modernization of the medieval university, the process was powerfully inflected by the rise of princely academies in the 16TH and 17TH centuries, and, in their wake, of learned societies and national academies in the 18TH and 19TH centuries. Each of these had their own licenses on knowledge, as well as professional rituals, meetings, and publications. By the second half of the 19TH century, with industrialization in full swing and the building of public school systems and public universities underway in Europe and the United States, the humanities began to assume their contemporary guise. This is the era in which the study of literature, philosophy, and classics was split off from the natural and physical sciences, even as "history" and the historical disciplines came to be understood as expressions of *Wissenschaft* in the double sense of a "science" and a discipline endowed with a distinctive toolkit for grappling with the cultural record.

Within this universe, the edifice of the humanities was firmly anchored in classical philology with fields such as archaeology, art history, and linguistics emerging only gradually from the shadow of textual studies. At the turn of the 20TH century, in the Anglophone context, departments of literature began to be established as departments of, first, medieval and Renaissance and, later, modern philology. Focused primarily on the study of language and rhetoric, they soon became organized by national literature groupings and by media. Though their roots reach back to Goethian notions of *Weltliteratur* and to 19TH century departments of comparative philology, comparative literature departments begin to emerge on the worldwide stage during the interwar period, in the midst of political upheaval, resurgent nationalism, and the threat of totalitarianism. A key moment in this history is marked by the post-World War II diaspora that saw classically trained philologists such as Erich Auerbach, Leo Spitzer, and René Wellek cross the Atlantic to take up positions in leading American universities.

Significantly, text-based disciplines and studies (classics, literature, philosophy, the history of ideas), make up, from the very start, the core of both the humanities and of the Great Books curricula instituted in the 1920s and 1930s. (For all their importance to the history of civilization, "Great Dance Performances" or "Great Architecture" never formed the basis of liberal arts curricula.) In other words, modern concepts of humanistic knowledge were built on authoring, narrative, and textual models specific to the medium of print, with the monograph gradually supplanting commentaries and critical editions as the inviolable touchstone of scholarly knowledge and achievement. Such models were, and still are, deemed to provide essential skills in rhetoric and analysis considered crucial in training for the professions of law, clergy, military, and statesmanship. By the mid-20TH century, the modern research university assumed its present form, with segmented humanities departments separated from the natural and social sciences as well as from vocational and professional schools. Digital work challenges many of these separations, promoting dialogue not only across established disciplinary lines but also across the pure/applied, qualitative/quantitative, and theoretical/practical divides.

But to make the argument for why the humanities remain more necessary than ever, we have to go beyond mere bromides celebrating the inherent value of cultural tradition or the inherent value of a familiarity with certain achievements from the cultural-historical past. No matter how imperiled by vocationalism, cost-cutting administrators, or the self-inflicted wounds of internecine battles, the humanities must survive because they embody distinctive modes of producing knowledge and distinctive models of knowledge itself. We refuse to take the default position that the humanities are in "crisis," in part because this very rhetoric of crisis has persisted for well over a century, however many mutations it has undergone. Jeremiads regarding the decline of educational standards, the failure of students and faculty alike to adequately embrace humanistic ideals, and the demise of tradition may well be inherent to the process of education itself. *Digital_Humanities* adopts a different view: It envisages the present era as one of exceptional promise for the renewal of humanistic scholarship and sets out to demonstrate the contributions of contemporary humanities scholarship to new modes of knowledge formation enabled by networked, digital environments.

Beginnings of Digitization

The first waves of the humanities' engagement with networks and computation embraced pioneering work from the late 1940s and the models that inspired archival projects at Oxford in the early 1970s. Over subsequent decades, the humanities continued to imagine the digital as a way of extending the toolkits of traditional scholarship and opening up archives and databases to wider audiences of users. These activities typically focused on corpus building, on creating standards for text encoding, and on building databases that could facilitate work on humanistic corpora, as librarians and information specialists developed machine-readable records, file formats, and systems that could support these ventures.

Gathering momentum from the late 1980s through the start of the 21ST century, a first wave of Digital Humanities developed, critiqued, and disseminated ways of structuring humanities data to dialogue effectively with computation. Database tools provided the foundation of the first Digital Humanities projects that were seeded around the world. Though this work was varied in nature, there were common, salient features: a concern with textual analysis and cataloging, the study of linguistic features, an emphasis on pedagogical supports and learning environments, and research questions driven by analyzing structured data. The migration of materials into digital forms and the extension of traditional methods of editing and analysis, enhanced by automation, took precedence. Important initiatives included the Perseus project, which converted the corpus of classical literature into digital form; the Women Writers Project, which created archives in which famous and obscure writers could coexist alongside an apparatus of cross references to their publications and textual borrowings; and The Valley of the Shadow, which posed questions about the role of primary documents in the work of cultural historians. Scholars then expanded and began to devise collaborative, multi-authored, cross-platform work on topics within their areas of specialization as well as to engage with emerging forms of digital culture. In this they were like the contemporary artists, poets, and musicians making imaginative use of algorithms to generate new works and taking advantage of communications networks to craft telematic projects or works in cross-media formats.

The advent of the Web in the early 1990s accelerated the transition in digital scholarship from processing to networking. The need for standards and conventions took on urgency, just as the need for a uniform gauge of rail or a point-size system for the casting of metal type or a common telegraphic code had in earlier moments of technological development. The graphical user interface introduced new possibilities and expectations. Games, entertainment, and immersive virtual environments all migrated online. Expectations about the quality of graphics rose as bandwidth opportunities exploded. The development of innovative, multimedia expressions of humanistic research in digital environments had to mature alongside these advancements. New tools and methods, new ways of thinking and working—what might be called "theory in practice"—all needed time to move beyond text-based models and immerse themselves in the multidimensional world of the Web. Scholars began to wrestle with the methods of mass-media art, corporate platforms, and entertainment, wondering if they should ignore them, make use of them, or counter them. The struggle is still underway.

In the late 1990s, projects began to appear that harnessed the digital to create visualizations, geospatial representations, simulated spaces, and network analyses of complex systems. For example, repository development on a massive scale, such as that undertaken by Europeana, engages multiple partners and stakeholders to make cultural legacy available to broad publics for a wide range of purposes. Questions about how to infuse the technological underpinnings of these approaches with humanistic methods and values remain. Challenges lie everywhere and, with them, opportunities to once again make explicit the value of humanistic modes of inquiry, thought, and creativity. How might the history of ancient scroll design and late medieval page layouts reshape our imaginings of the expressive possibilities of digital scrolling or digital page units? Can computational and digital environments be designed to capture the fluidity of an intercultural dialogue between diasporic peoples? What lessons can be carried over from successful forms of interactive media into the world of teaching or into the communication of research and historical knowledge to the public-at-large? What media forms and modalities of engagement might a critical edition of an audio recording assume? We see such questions and the many others that accompany them as harbingers of renewal, signs that this is a galvanizing moment to be a humanist involved in devising, designing, and deploying new tools; in opening expanded modes of inquiry unthinkable under pre-digital conditions; and in forging innovative, multimodal approaches to traditional questions (about authorship, influence, dissemination patterns) through the as-yet-unrealized possibilities of digital platforms.

Transmedia Modes of Argumentation

Printed books and humanistic scholarship have a shared history. For centuries, humanists have worked with formats—the printed page, the bound codex—that have remained essentially consistent. But communication in digital environments has required the invention of new forms, tools, and schemata. The lack of conventions and the opportunity to imagine formats with very different affordances than print have not only brought about recognition of the socio-cultural construction and cognitive implications of standard print formats, but have also highlighted the role of design in communication. Modeling knowledge in digital environments requires the perspectives of humanists, designers, and technologists.

In the 21ST century, we communicate in media significantly more varied, extensible, and multiplicative than linear text. From scalable databases to information visualizations, from video lectures to multiuser virtual platforms, serious content and rigorous argumentation take shape across multiple platforms and media. The best Digital Humanities pedagogy and research projects train students both in "reading" and "writing" these emergent rhetorics and in understanding how they reshape and remodel humanistic knowledge. This means developing critically informed literacies expansive enough to include graphic design, visual narrative, time-based media, and the development of interfaces (rather than the rote acceptance of them as off-the-shelf products). The second half of the 20TH century saw the development of such literacies in fits and starts. They now move front and center inasmuch as the advent of Digital Humanities implies a reinterpretation of the humanities as a generative enterprise: one in which students and faculty alike are *making things* as they study and perform research, generating not just texts (in the form of analysis, commentary, narration, critique) but also images, interactions, cross-media corpora, software, and platforms.

Because Digital Humanities is a generative practice, it demands an additive pedagogy. Students still have to be trained in the persuasive use of language, to write effectively in long forms, but they also need to be able to craft what Roman rhetoricians called the *multum in parvo*—the aphorism, the short form, that which distills the long and the large into compact form. This is not only to address the (perhaps apocryphal) short length of the contemporary attention span—was there ever a golden age of rapt audiences with limitless patience? rhetorical treatises from classical antiquity suggest that there wasn't—but also the realities of a wired world in which the "real estate" available for text and images is ever-shifting and in which argumentation must be able to expand and contract as a function of shifting constraints and technological affordances. Roman teachers of rhetoric would have no difficulty in understanding this challenge, but they might well wonder about our devaluation of the oral component of their ancient art. In the era of pervasive personal broadcasting, the art of oratory must be rediscovered. This is because digital networks and media have brought orality back into the mainstream of argumentation after a half-millennium in which it was mostly cast in a supporting role vis-à-vis print. YouTube lectures, podcasts, audio books, and the ubiquity of what is sometimes referred to as "demo culture" in the Digital Humanities all contribute to the resurgence of voice, of gesture, of extemporaneous speaking, of embodied *performances* of argument. But unlike in the past, such performances can be recorded, disseminated, and remixed, thereby becoming units of polymorphous exchange and productive mutation.

Digital Humanities necessarily partakes in and contributes to the "screen culture" of the 21ST century. From stationary computer monitors to mobile tablets, postage stamp sized-LCDs on communication devices to dynamic, building-sized imagescapes, screens have become pervasive in contemporary life. What this means is that the visual becomes ever more fundamental to the Digital Humanities, in ways that complement, enhance, and sometimes are in tension with the textual. There is no either/or, no simple interchangeability between language and the visual, no strict subordination of the one to the other. Words are themselves visual but other kinds of visual constructs do different things. The question is how to use each to its best effect and to devise meaningful intertwinglings, to use Theodor Nelson's ludic neologism. The visual does not necessarily represent an advance over the capabilities of text. It is simply a different, distinct medium for thinking, communicating, and working, with its own rigors and histories, its own skill-sets and language, and its own freedoms and constraints.

The suite of expressive forms now encompasses the use of sound, motion graphics, animation, screen capture, video, audio, and the appropriation and remixing of code that underlies game engines. This expanded range of communicative tools requires those who are engaged in Digital Humanities work to familiarize themselves with issues, discussions, and debates in design fields, especially communication and interaction design. Like their print predecessors, format conventions in screen environments can become naturalized all too quickly, with the result that the thinking that informed their design goes unperceived. Though there is no "natural" way to interweave text, images, sound and moving images, there exists a range of available genre models from experiments unique to the digital realm to ones that draw upon prior moments in the history of print and cinematic conventions. Digital design expresses concepts by means of the multitude of ways in which it layers media, structures information, and articulates navigational strategies. Though not every project requires a custom approach or platform, attention to the design of arguments is a fundamental feature of Digital Humanities research.

Designing Digital Humanities

Like the word "writing," the word "design" encompasses an array of activities from the everyday to the highly specialized. "Big D" design ranges from the business plans and systems of "design thinking" to the "design sciences," which include engineering and human-computer interaction, to the cultural critique and speculative provocations of "critical design." In between are myriad professional specializations and academic domains. Digital Humanities projects most closely involve communication/graphic/visual designers who are concerned with the symbolic representation of language, the graphical expression of concepts, and questions of style and identity. Interaction/user experience designers, with their focus on interface, behavior, and digital systems, and media designers who combine communication and interaction also bring expertise that is critical to the design of operations and environments that structure the ways in which ideas come into being.

In generative mode, these designers shape structural logics, rhetorical schemata, information hierarchies, experiential qualities, cultural positioning, and narrative strategies. When working analytically, their task is to visually interpret, remap or reframe, reveal patterns, deconstruct, reconstruct, situate, and critique.

To design new structures of argumentation is an entirely different activity than to form argumentation within existing structures that have been codified and variously naturalized. All forms of design share a propositional orientation that is well-suited to the challenges that come with designing new structures, for design asks: "What if?" Each design iteration plays out an answer to the question: "What happens when...?" In a world with fluid contours, humanists, designers and technologists working together can move beyond considering what can be done with the tools at hand to ask: "What can we imagine doing that may not yet be possible?"

For digital humanists, DESIGN is a creative practice harnessing cultural, social, economic, and technological constraints in order to bring systems and objects into the world. Design in dialogue with research is simply a technique, but when used to pose and frame questions about knowledge, design becomes an intellectual method. In the hundred-plus years during which a self-conscious practice of design has existed, the field has successfully exploited technology for cultural production, either as useful design technologies in and of themselves, or by shaping the culture's technological imaginary. As Digital Humanities both shapes and interprets this imaginary, its engagement with design as a method of thinking-through-practice is indispensable. Digital Humanities is a production-based endeavor in which theoretical issues get tested in the design of implementations, and implementations are loci of theoretical reflection and elaboration.

In addition to modeling the platforms, tools, databases, and other information structures on which digital projects are built, designers understand the possibilities and limitations of each of the specific media forms employed in such projects. Digital humanists have much to learn from communication and media design about how to juxtapose and integrate words and images, create hierarchies of reading, forge pathways of understanding, deploy grids and templates to best effect, and develop navigational schemata that guide and produce meaningful interactions. Not every digital humanist will become a designer, but every good digital humanist has to be able to "read" and appreciate that which design has to offer, to build the shared vocabulary and mutual respect that can lead to fruitful collaborations. Understanding the rhetoric of design, its persuasive force and central role in the shaping of arguments, is a critical tool for digital work in all disciplines. But rhetoric is a distinctly humanist skill, one that ventures out into new directions in a digital environment that the humanist of the 21ST century is called upon to master.

A number of influential 20TH century media culture experiments that combine the visual and the verbal in equal measure provide a glimpse at the potential of collaborations between design and the humanities. The confluence of Marshall McLuhan's words with Quentin Fiore's images and book design in *The Medium is the Massage* could be seen as a precursor to contemporary Digital Humanities work, both for the form of its argument and for its collaborative production, orchestrated by producer Jerome Agel. Similarly, John Berger's book *Ways of Seeing* is meant to be both viewed and read in what could be considered a prototypical transmedia project: The book was originally a BBC television series. And while graphic novels and comics are by definition a combination of words and pictures, Scott McCloud's *Understanding Comics* is a noteworthy graphic nonfiction essay: it enacts an analysis of the interplay between text and image in spatialized sequential narratives through the use of text and image in a spatialized sequential narrative.

Each of these projects brought new forms of argumentation to the static page. But the screen culture of Digital Humanities is often dynamic and time-based, drawing on a multitude of traditions of media practice. Here, the aesthetics and technics of film and video are particularly relevant. Being able to block out sequences and actions, light and frame shots, edit for sense and rhythm, and compose and produce music and sound—this and more comprise the fundamentals of moving image production. Techniques for editing shots to create scenes, narratives, or emotional effects, mixing in sound, virtual simulations, and other special effects to create a cohesive whole are the essence of what is referred to as "post-production." One need only consider the subtle and tightly controlled interplay among words, sound, and images of films such as Chris Marker's *Sans Soleil* or Errol Morris' *Fast Cheap & Out of Control* to understand that these techniques are—as with design—about more than simply production: They are the means with which to investigate and articulate an idea.

The addition of other graphic supports such as charts, graphs, and animations, which are often essential in making a Digital Humanities argument, tend to extend the process even beyond the classical structures of film and television aesthetics into the hybridized realms of the motion graphic or information visualization. An early example of this mixing is Charles and Ray Eames's short film *Powers of Ten* which combined a filmic first person perspective with didactic narration and information graphics to create a complete work whose sum is greater than its parts—what designers refer to as the gestalt. Now, distributed digital systems make it possible to combine live data streams and interactive systems in which real-time input can be displayed on maps, projection systems, and immersive 3-D environments, animated by means of a rich array of "born digital" visual effects. Processing embedded sensor input or engaging with feeds from social media challenges the very concept of the archive which has now come to encompass the realm of live, unfolding events. The design of each of these dynamic aspects is not simply a display or interface "problem" to be "solved"—it is, as with *Powers of Ten*, the embodiment of a project's argument and methodology.

Digital media have become the meta-medium par excellence, able to absorb and re-mediate all previous forms in a fluid environment in which remixing and culture jamming are the common currency. In the realm of Digital Humanities practice, designing the cultural record is an act of thinking, and design processes become multivalent. This openness, an outgrowth of the iterative and (almost) infinitely mutable and expansive nature of digital media, stands in contrast to inherited notions of "writing" or "picture-making" or "printing"—all of which are stabilizing practices with slow refresh rates. If texts in their broadest sense can be thought of as "media scripts," then the specific medium that instantiates that script can change, evolve, morph, and even turn back upon itself. The rewritable substrate of digital media enables iterative work to hitherto unprecedented degrees, introducing the software term "version" into units of scholarly production.

The field of Digital Humanities may see the emergence of polymaths who can "do it all": who can research, write, shoot, edit, code, model, design, network, and dialogue with users. But there is also ample room for specialization and, particularly, for collaboration. The generation now cursed with the label "digital natives" will surely develop the capacity to become comprehensive digital humanists. The fact that digital projects of any substantial scale benefit from and, indeed, often require team-based approaches troubles traditional concepts of authorship in the humanities, which are still fixated, by and large, on single-authored achievements. The academic world has developed sophisticated (though hardly perfect) modes in the sciences to credit multiple authors, but colleges and universities now need to develop ways of acknowledging intellectual contributions in team environments for digital humanists, a micro-credit and a macro-credit system for intellectual labor that functions as a viable form of capital in a reputation economy as well as in a scholarly world. Technical imagination and expertise partner with discipline-specific forms of knowledge in Digital Humanities projects: projects in which each contributor plays a vital role in setting the research agenda, and in which contributors build big mosaics out of tesserae consisting of specialized skills and expert knowledge.

One caveat is worth noting. The positive demand for expanded skill-sets could have profoundly negative effects on scholarship if it becomes the academic equivalent of a neo-liberal speedup in which ever more quantitative metrics are used to push "education workers" into acquiring technological skills without commensurate pay, skills which they are then held accountable for, both within and outside of tenure tracks. Likewise, the continuing resistance within post-secondary educational institutions to recognize Digital Humanities work as equivalent to long-established forms of scholarship could translate into an expectation that certain disciplines devoted to the study of the contemporary, such as media and visual studies, become Digital Humanities departments, irrespective of whether the most promising research questions within the field are well-suited to such a framing. The fact of the matter is that Digital Humanities bears no privileged relation to modern or contemporary cultural corpora; on the contrary, it is indifferent as to whether its objects of study are performance videos from the 1960s or pottery shards from a Mycenaean archaeological site from the 2ND millennium BCE. Digital Humanities is an extension of traditional knowledge skills and methods, not a replacement for them. Its distinctive contributions do not obliterate the insights of the past, but add and supplement the humanities' long-standing commitment to scholarly interpretation, informed research, structured argument, and dialogue within communities of practice.

In this rapidly changing research environment, it is necessary to acknowledge the new shapes that knowledge production is assuming, to set reasonable and flexible expectations regarding experimentation and innovation, and to devise a reward structure for team-based collaboration that includes recognition of the value and skills of participants in accord with the significance of their contributions. Older "service-based" models of staff conceived in contrast to scholars qua *auteurs* are being challenged and rightly so. The cultural politics of academic institutions are shifting, indeed, but we must be attentive to inadvertent consequences. Projects that are dependent on deliverables as their only measure of success are likely to be at odds with a research mission that supports innovation and imaginative, risk-taking work. Intellectual challenges, not technical ones or metrics based on the mere on-time delivery of products, have always driven and will continue to drive the development of the Digital Humanities.

Computational Activities in Digital Humanities

Digital Humanities projects can be described by sketching their structure at several levels. These begin with basic computation (programming, processing, protocols) and extend through the levels of organization and output that form the basis of most users' experience (interface, devices, networks). The foundational layer, COMPUTATION, relies on principles that are, on the surface, at odds with humanistic methods. Specifically, computation depends on disambiguation at every level, from encoding to the structuring of information. Explicit step-by-step procedures form the basis of computational activity. However, ambiguity and implicit assumptions are crucial to the humanities. In the intersection between these two domains, humanists have given in to the demands of a process that requires that they work in accord with its methods. What is less-often noted is that computational methods have been altered in significant ways by humanist approaches. Indeed, this is a challenge for the development of the Digital Humanities, namely the ways in which ambiguity, interpretation, contingency, positionality, and differential approaches can be embodied in computation.

The second level involves PROCESSING in ways that conform to computational capacities, and these were explored in the first generation of digital scholarship in stylometrics, concordance development, and indexing. This processing activity takes advantage of the ability of computers to automate certain tasks useful in answering the sorts of research questions that were initially being asked by humanities scholars. In the first phase of digital activity, sorting, searching, calculating, and matching were basic operations performed on texts or data. The introduction of structured data for analysis and display in the family of what are known as markup languages added a dimension to this activity, introducing interpretation into the digitized stream of keyboarded characters. The insertion of these "tags" allowed manipulation of the content and the performance of an interpretive act.

Both computational foundations and processing activities have endured, but other platforms, tools, and infrastructures have also developed to support curation, analysis, editing, and modeling. These depend upon the basic building blocks of digital activity: DIGITIZATION, CLASSIFICATION, DESCRIPTION and METADATA, ORGANIZATION, and NAVIGATION. Designing and building digital projects depend on knowledge of these fundamentals and on a nuanced understanding of the networked environments in which the projects will develop and variously reside.

Curation, Analysis, Editing, Modeling

Curation, analysis, editing, and modeling comprise fundamental activities at the core of Digital Humanities. Involving archives, collections, repositories, and other aggregations of materials, CURATION is the selection and organization of materials in an interpretive framework, argument, or exhibit. The capacity with digital media to create enhanced forms of curation brings humanistic values into play in ways that were difficult to achieve in traditional museum or library settings. Rather than being viewed as autonomous or self-evident, artifacts can be seen being shaped by and shaping complex networks of influence, production, dissemination, and reception, animated by multilayered debates and historical forces.

ANALYSIS refers to the processing of text or data: Statistical and quantitative methods of analysis have brought close reading of texts (stylometrics and genre analysis, collation, comparison of versions for author attribution or usage patterns) into dialogue with distant reading (the crunching of large quantities of information across a corpus of textual data or its metadata). Analysis is often conjugated with visualization in order to give graphical legibility to analytical results. Many of the tools for visualization are still adopted wholesale from business graphics or from the social and natural sciences, but this is beginning to change as data visualization assumes an evermore central role in Digital Humanities scholarship. The recent surge of interest among digital humanists in mapping, for example, is indicative of a trend that recognizes the importance of developing geo-temporal visualizations and mapping platforms to analyze complex social, cultural, and historical dynamics.

EDITING has been revived with the advent of digital media and the Web, and will continue to be an integral activity in textual as well as time-based formats. The parsing of the cultural record in terms of questions of authenticity, origin, transmission, or production is one of the foundation stones of humanistic scholarship upon which all other interpretive work depends. But editing is also productive and generative, and it is the suite of rhetorical devices that make a work. Editing is the creative, imaginative activity of making, and as such, design can be also seen as a kind of editing: It is the means by which an argument takes shape and is given form. Genetic editions, in which variants, versions, pentimenti, and amendments can be incorporated into a display or trail of evidence have been the dream of literary scholars since the rise of scientific philology in the 19TH century. Tools for the realization of such complex forms of intellectual gamesmanship are changing and improving rapidly. The potential for their full realization even beyond the confines of the textual record will revitalize long-standing traditions of humanistic work and allow humanists to re-approach these traditions in innovative ways with new research questions and tools.

MODELING highlights the notion of content models—shapes of argument expressed in information structures and their design. A digital project is always an expression of assumptions about knowledge: usually domain-specific knowledge given an explicit form by the model in which it is designed. Thus a project dedicated to analyzing the correspondence of a famous artist might assume a chronological shape, which is one model of a human life. Or it might be organized around correspondents and relationships, another way of weighting the data. Or it could be structured by place of origin and receipt, as a geospatial network. The building blocks of digital work will each be molded by the model of knowledge which they need to serve. Even basic questions about file formats, image resolution, metadata, and classification schemes to structure the digital materials are intimately bound to the argument made by what is referred to here as a "content model." The phrase means just what it appears to mean: a model by means of which shape is conferred upon a given set of cultural contents. Do we organize music files by playlists or by artist? By performer or composer? The playlist model fixes files in an order that makes searching for a particular artist difficult, and classical music might be more logically organized by composer than by performing artist. Each of these represents a distinctive information model that privileges one or another feature of the content.

The organization of information in a file or data system does not have to conform to its display within an interface. At the level of interface, one might well create a design that is based on the behaviors that end-users might plausibly display with respect to the information. Do they want to search (look for a particular thing) or browse (wander about in a collection to see what might be of interest)? Such distinctions are the bedrock upon which interaction design is built. The knowledge for carrying out the implementation of these designs comes from computer science, information studies, graphic and media design, human-computer-interaction, and cognitive studies. The form that knowledge takes in digital environments and the arguments it expresses in its information structures can be deeply infused with humanistic values, but only if humanists are involved. If simply handed off to technologists or left to functionaries or commercial interests, many basic requirements for humanist scholarship and pedagogy will be lost. The misguided collector annoyed by the mass of handwritten annotations created by readers in the margins of medieval manuscripts and incunabula who elects to erase them eliminates forever the commentary of famous and insignificant figures alike. In a digital world, choices about what remains and what is eliminated, what is made accessible, how and in what form, are just as enduring and just as potentially enhancing or damaging.

Additionally, modeling carries a specific meaning in the creation of simulated and virtual environments. Rendering immersive models of historical sites, archaeological projects, cultural monuments, or imagined worlds in fly-through, multidimensional forms are vivid possibilities of the digital environment. So are the multiplayer worlds of games in which participants make virtual real estate and its contents, creating systems of value, social relations, and lived experiences in avatar-inhabited landscapes. Humanities work in such environments allows questions of uncertainty and analysis to enter into play. The role of speculation in the use of fragmentary evidence mustered for virtual reconstruction gets amplified through the capacities of digital media. Digital humanists engage with these environments not only because of their pedagogical and research values, but also because humanist sensibilities are needed to challenge the seductive force of seamless presentation and to inject criticality and skeptical faculties into otherwise "naturalized" unnatural constructs.

The graphical user interface, still common in a world of distributed and embedded computing platforms, has put tremendous pressure on this generation of scholars and teachers to be attuned to sophisticated visual literacy. Even the most text-centric academic will admit the existence of visual rhetoric, but the skills to read interfaces, databases, and other content models are still very underdeveloped. Understanding the way one structures the relationships among data, the ways in which users input and access information, and the physical and conceptual design of such systems all-too-often slips away into the abstraction. Yet graphical interfaces have been central to the humanities for centuries. What, after all, are indexes, tables of contents, and foot- and endnotes if not information storage and retrieval strategies? The classification systems that scholars and librarians have evolved over the centuries and their direct relationship to the arrangement of physical book stacks, not to mention whether those stacks are open or closed, are all evidence of the design of information and its access as central concerns of the humanities. Yet with computers and networks, these same issues of information and access may be perceived as mere technical concerns, and the benefit of a humanist perspective is lost. Navigation and organization are interdependent; creating digital wayfinding, like environmental signage, calls on a combination of intellectual and design skills.

Each of these areas of activity—curation, analysis, editing, and modeling—is supported by the basic building blocks of digital activity. But they also depend upon NETWORKS and INFRASTRUCTURE that are cultural and institutional as well as technical. Servers, software, and systems administration are key elements of any project design. Compatibility and interoperability are essential for sustainable work. The cultural dimensions of infrastructure are also factors to be considered. Museums, libraries, archives, and other institutional settings each have their own agendas, their own customs and conventions. Cultural differences can arise with partnering institutions, as well as across national and international communities of participants. Digital work takes place in the real world, and humanists once accustomed to isolated or individualized modes of production must now grapple with complex partnerships and with insuring the long-term availability and viability of their scholarship.

Prototyping and Versioning: Generative Humanities Ahead

The capacity for the rapid creation, testing, and reworking of Digital Humanities projects goes hand-in-hand with the flexibility, mutability, and extensibility of digital media. But with the development of more Digital Humanities projects comes a new, normative center in which tool sets are stabilizing. Curation, collection, and data management are cohering around shared standards, while concrete rationales for the production and deployment of Digital Humanities methodologies have emerged in the academy. This normalization points, in part, to the maturation of the Digital Humanities. However, one of the strongest attributes of the field is that the iterative VERSIONING of digital projects fosters experimentation, risk-taking, redefinition, and sometimes failure. It is important that we do not short-circuit this experimental process in the rush to normalize practices, standardize methodologies, and define evaluative metrics.

Whereas the first generation of Digital Humanities tended to specialize in discrete one-offs, digital humanists can now use networks and interoperable file-sharing standards and protocols to test new approaches with distributed users and developers at a time-and-distance scale previously unimaginable. Digital Humanities infrastructures encourage PROTOTYPING, generating new projects, beta-testing them with audiences both sympathetic and skeptical, and then actually looking at the results. Building on a key aspect of design innovation, Digital Humanities must have, and even encourages, FAILURES. Outside the normative core, there is space to iterate and test, to create precarious experiments that are speculative, ludic, or even impossible. That research can benefit from failure should not be any sort of surprise—stress-testing metals and other materials is what gives us bridges that don't collapse and buildings that stay up—but so too can the classroom benefit from an academic culture that welcomes frequent (productive) failure. The methodologies of Digital Humanities are robust precisely because they place lasting pedagogical value in the creative, generative, and experimental processes of design-based research. Process is favored over product; versioning and extensibility are favored over definitive editions and research silos. The Digital Humanities capacity to ask, design, and model new research questions opens new possibilities for those who are willing to take risks. Too often in established cultural discourse, the experimental is absent or hastily erased, the dialogue already so well-established that new approaches are incremental at best. But it is not an experiment if it cannot fail.

Many of the most promising areas of the Digital Humanities have ample room for such risky undertakings. The key is to create the contexts that allow failing to be seen as something other than defeat. In the entrepreneurial culture of Silicon Valley, for example, failure is not only tolerated, it is massively funded—because the risks are worth it. Industry leaders factor the costs of failure into labor, resources, talent, and investment as a necessary part of their undertakings, recognizing the need for experimentation with uncertain outcomes. As Bill Coleman, who has had many wins but even more losses over the decades in the high-tech industry, notes, "You learn not just about failure and how to make things work, you learn the psychology of failure and how you react to it."

Accepting the psychology of failure is part of the life cycle of innovation. Yet when the academic culture of peer review and promotion runs up against budget realities and resource scarcity, skittishness about failure arises. Digital Humanities work embraces the iterative, in which experiments are run over time and become objects open to constant revision. Critical design discourse is moving away from a strict problem-solving approach that seeks to find a final answer: Each new design opens up new problems and—productively—creates new questions. Digital humanists take these matters as core tenets, knowing that the field we are engaged in is still in its early stages, both enabled and constrained by encounters with the techniques of computation. We are driven by the need to extend the reach and impact of that technology to forge ahead; when we arrive at the place where humanistic methods regularly inform computational approaches, we will have passed another exponential milestone.

Generative Humanities as the New Core

The uptake of digital tools and platforms does not "solve" long-standing problems in the academy as much as offer what engineers refer to as "work-arounds," provisional improvisations that allow whole systems to move forward without demanding perfection from every part. The iterative nature of the Digital Humanities is what makes this a once-in-a-generation moment to reinvigorate the idea of a core curriculum for undergraduates: to make them active participants and stakeholders in the creation and preservation of cultural materials. Why would the Digital Humanities want to wade into what feels like a never-ending academic culture war? In the United States in particular, we have never settled on what constitutes the "basic" things an educated person should know, and how that knowledge in turn develops the informed citizenry a democracy needs to thrive.

The last 50 years saw the growth of increasing discomfort with inherited curricula, which were rightly seen as constrained by issues of race, class, gender, and first-world biases rooted in Eurocentric traditions. An important battle took place—to open reading lists and discussion sections to a wider range of voices. Yet this call for openness and expansion dovetailed with the silo-ization of knowledge in the humanities as the baby boom generation hit the newly expanded higher education sector in the 1960s. Students clamored for relevance; activists demanded inclusion; and scholars responded by opening up their syllabi while at the same time narrowing their teaching to reflect and feed their specializations. Figures and movements formerly ignored precisely because of their supposedly "marginal" status became new objects of study. Perhaps more significantly, the perspectives of these once-excluded materials carried with them alterative methodologies and different value systems that shattered any illusion of a single belief system within humanistic thought.

The wars over the core have had two unexpected results. The first is that rather than replacing a restrictive body of knowledge with a more expansive one, the very idea of sharing common references or approaches waned. The wars over the core in the humanities have contributed to a malaise in which the humanities are widely perceived as "irrelevant," lacking the practicality of business, law, or medicine. Another effect has been to add ammunition to the forces that want to de-college the American populace, shunting as many students as possible into vocational tracks, in order to reserve higher education for elites. Yet the reality is that graduates of whatever level will need to call upon more than vocational training if they are to steer their democracy through the challenges and opportunities that this highly networked, globalized, mobile, and ecologically fragile century offers. More than ever, we need the critical insights, creative designs, speculative imagination, and methods of comparative, historically informed study that shape humanistic modes of inquiry. Imagination and informed critical thought foster ways of thinking beyond received positions and claims to absolute authority. Digital, polyvocal expression can support a genuine multiverse in which no single point of view can claim the center. The principles of relativist approaches to knowledge, rooted in historically situated understanding, remain fundamental to (digital) humanism.

The phrase *Digital Humanities* thus describes not just a collective singular but also the humanities in the plural, able to address and engage disparate subject matters across media, language, location, and history. But, however heterogeneous, the Digital Humanities is unified by its emphasis on making, connecting, interpreting, and collaborating. This concentration on process and method might in fact be the way to develop a work-around for the creation of a core curriculum, a process which bogs down precisely on what appears to its varied partisans to be a zero-sum game. An Afro-Caribbean female novelist joining the syllabus means an English male metaphysical poet exits. In the eight semesters of the hypothetical student's college career, there are only so many class sessions. But the networked academy's very *allatonceness*—to use Marshall McLuhan's suggestive term referring to simultaneity and connectivity—offers a glimpse of a more elastic notion of curricula, one that extends past the walls of the academy and the time limits of degree programs. At the very same time that the battles over the core raged on, the entertainment and information industries flourished. The disconnect between methods of pedagogy inherited from cloisters and seminar rooms and those of a massively mediated culture is real. Digital humanists strive to bridge that gap.

The digital environment offers expanded possibilities for exploring multiple approaches to what constitutes knowledge and what methods qualify as valid for its production. This implies that the 8-page essay and the 25-page research paper will have to make room for the game design, the multi-player narrative, the video mash-up, the online exhibit and other new forms and formats as pedagogical exercises. Playful, imaginative, participatory work is not the enemy of education, but its exuberant and vital engine. New standards of assessment will be necessary as skills change. We struggle less to remember facts than we do to remember where and how to find them—and how to assess their validity.

Ubiquitous networks have led and will continue to lead to evolutions in pedagogy precisely because they involve the outsourcing of memory. Writing transformed traditional modes of oral training; print technologies standardized everything from spelling to what constituted a "proper" copy of a text. We would be ignoring precedent completely if we assumed that the *allatonceness* of a vast and increasing digital archive accessible anywhere at any time will not affect the way that we learn. The best core curricula—whether or not based on classical models—strive to create students, and thereby citizens, who think with imagination, who manifest their thoughts as creative action, and whose analysis can lead to inventive, although hardly definitive, syntheses. These are precisely the goals that a digitally driven, generative humanities core espouses.

The technological aspects of the digital turn are not yet so normative that we can ignore the tools, interfaces, and the hard-, soft-, and wet-wares of this moment. But the generative humanities are emphatically not about training for a market. They are, instead, like all great pedagogies that preceded them, education for an environment. The social, political, and ecological challenges of the 21ST century demand significantly more than textual analysis or recitations of inherited content. These problems (and opportunities) will need people trained to create synthetic responses, rich with meaning and purpose, and capable of communicating in a range of appropriate media, including but not limited to print. The exact content of the generative humanities qua core curriculum will always be a matter of negotiation and debate; and well it should be, for core curricula have always been in greater flux than their adversaries or diehard advocates care to admit. Some traditionalists will rankle at the idea that the humanist spirit—rather than humanities texts—will become the "core" of a generative humanities curriculum, but this century's explosion of a deep, rich, and meaningful digital culture is already proving them wrong. That spirit, as suggested throughout this chapter, consists of methods as well as content, with approaches that tolerate relativism and diversity in thinking, orders of experience, and, yes, fundamental values.

The generative aspects of Digital Humanities thus go a long way to addressing the much-lamented atomization and irrelevance of scholarship—that critique from all parts of the ideological spectrum that teaching and research are at odds with one another, that scholarship itself has become relentlessly focused on the professional advancement of the scholar and is addressed only to others in an ever-shrinking pool of the like-minded and credentialed. Digital Humanities scholarship, on the other hand, promises to expand the constituency of serious scholarship and engage in a dialogue with the world at large. Even as it models ever-newer forms of professional expertise, Digital Humanities employs the best crowd-sourcing techniques to process, analyze, and publish materials that document and engage with the variance of the human cultural record. It promotes platforms for informed amateur scholarship, and it serves to make humanities research into something of a new multi-player online game with global reach and relevance. In its distributed form, Digital Humanities arrives through cellphone and other mobile applications as a deepening enhancement of daily experience, providing an interpretation of a public monument or work of cultural legacy, bringing the richness of scholarly expertise into new and decidedly public forms of use. In the world of current events and unfolding occurrences in the political or cultural sphere, rapid communication on digital platforms alters perception, opinion, values, and outcomes.

The digitization of the world's knowledge and its movement across global networks, no matter how incomplete or incompletely free, have transformed what we understand by and how we approach the humanities in the 21ST century. We are continually creating new ways of accessing and assessing this new cultural production, which continually open up important new spaces for exploring humanity's cultural heritage and for imagining future possibilities using the transmedia methods and genres of the digital present. It is to these methods and genres that we now turn.

2. EMERGING METHODS AND GENRES

HUMANITIES KNOWLEDGE USED TO HAVE A RECOGNIZABLE FORM.

WE KNEW IT WHEN WE SAW IT BECAUSE IT HAD LOOKED THE SAME FOR CENTURIES: PRINTED PAGES WITH LINEAR PROSE AND A BIBLIOGRAPHIC APPARATUS WRITTEN BY AN AUTHOR AND PUBLISHED IN THE FORM OF AN ARTICLE OR BOOK. THOUGH THE FORMAT COULD VARY WIDELY, FROM MATERIALS TO LAYOUT, SUCH DESIGN DECISIONS WERE RARELY CARRIED OUT BY THE SCHOLARS WHO CREATED THE CONTENT. WITH FEW EXCEPTIONS, THE HUMANITIES HAVE ADOPTED HOMOGENEOUS APPROACHES TO PRODUCING SCHOLARLY RESEARCH. YET ARTIFACTS CREATED BY DIGITAL TECHNOLOGIES THAT “LIVE” IN DIGITAL ENVIRONMENTS ARE COMPARATIVELY DIFFERENT—IN TERMS OF MATERIAL COMPOSITION, AUTHORSHIP, MEANING-MAKING, CIRCULATION, READING, VIEWING, NAVIGATION, EMBODIMENT, INTERACTIVITY, AND EXPRESSIVITY—FROM ARTIFACTS CREATED BY THE WORLD OF PRINT.

DIGITAL MEDIA are not more "evolved" than print media nor are books obsolete; but the multiplicity of media and the varied processes of mediation and remediation in the formation of cultural knowledge and humanistic inquiry require close attention. We strongly believe that humanists need to apply the same kind of rigorous media-specific, social, cultural, and economic analyses that we have honed to study print culture to understand the specificity and affordances of digital culture and to interrogate the status of knowledge, the concept of culture, and the redefinition of the social in our global information age. We also believe that humanists must actively engage with, design, create, critique, and, finally, hack the environments and technologies that facilitate this research as we render this world-as-a-world to help us produce knowledge about who we are, where we live, and what that means.

The purpose of this chapter is to provide a field map of the experimental forms and different "knowledge models" emerging in the Digital Humanities. This is not meant to be an exhaustive or definitive list of new methods and genres but rather a conceptual and theoretical introduction to emergent practices of scholarly inquiry. We move from examining the impact of technology on the most established of humanistic practices—the decisions about what constitutes a text and its variants—to positing that enhanced critical curation of those texts makes possible augmented editions and fluid textualities that rely on the affordances of digital environments. This fluidity allows digital humanists to play with scale, both in terms of how they approach data and how they model their results. Toggling between distant and close, macro and micro, and surface and depth becomes the norm. Here, we focus on the importance of visualization to the Digital Humanities before moving on to other, though often related, genres and methods such as locative investigation, thick mapping, animated archives, database documentaries, platform studies, and emergent practices like cultural analytics, data-mining, and humanities gaming. All of these are then situated within a technological matrix that almost demands the repurposing and remixing of cultural content. We conclude this chapter by considering the utopian prospect that the massive spread of shared knowledge across networks could give rise to a state of "ubiquitous scholarship," of ever-more interconnected, publicly engaged, participant citizens.

It is a tremendously exciting time for the humanities, as knowledge not only looks and sounds markedly different than it once did, but also feels different because it is experienced in new contexts and environments and created in collaborative spaces that involve communities who rarely, if ever, had the chance previously to participate in the scholarly enterprise.

The chapter is divided into emerging areas of experimentation, as represented in the index below. A Portfolio of Case Studies follows, providing concrete examples of these emerging genres and methods in application. Through linkages both graphical and conceptual, they combine to provide a lively and practical demonstration of Digital Humanities theory in practice.

ENHANCED CRITICAL CURATION

digital collections   multimedia critical editions   object-based argumentation   expanded publication   experiential and spatial   mixed physical and digital

Collection-building and curation have always defined humanistic learning: so much so that even the most ancient literary forms adopt listing, cataloging, and inventorying as key features of poetic communication. Inventories abound in Hesiod, Aeschylus, Sophocles, and Euripides. The epigrammatist Callimachus composed the 120 tablets of the *Pinakes* in which the entire holdings of the Alexandrian Library were enumerated. Apollonius of Rhodes opens the lead section of his *Argonautica* with a panoramic listing of the Argonauts. And the sheer proliferation of catalogs in the Homeric epics shows how cataloging can put into play a vivid mode of representation that is neither that of the laundry list with its skeletal seriatim organization nor that of an exhaustive, didactically intentioned inventory, but rather a composition that treads along the boundary line between verbal and visual figuration in the pursuit of informational concision and compactness. A poem within the poem, a condensation of names, actions, and things, the catalog is an art of memory that is also an art of data compression and of performance. In short, *katalégein* designates poetic composition as a compression algorithm and audience reception as a decompression tool some 26 centuries before the word "digital" began to refer to 0's and 1's.

Awareness of this historical background is crucial in order to understand how collection-building and curation have remained constants of humanistic knowledge production from remote antiquity through early modern courts to the academies of the Baroque era to late 19TH century universities where chairs were typically associated with the research collections. These domains became disjoined from the mainstream of scholarly practice only during the late print era, and are once again becoming integral to many forms of Digital Humanities practice.

The accumulation and care of knowledge were paramount within classical, premodern, and early modern regimes of data scarcity. In those eras, bits of information were valuable *a priori* and therefore either preserved, relayed, or reused, irrespective of whether they could be integrated into a cohesive structure or system of belief. Copiousness and copying were understood as inherently good under such circumstances, and collecting served the project of gathering, conservation, and retrieval. With the spread of print and the rise of modern institutions of memory (with their systematic approaches to collection and conservation), a new regime arises within which there exist such proliferations of historical information and cultural material that data from the past can no longer be assumed to possess *a priori* value. They become supports for the production of knowledge, knowledge's precondition but not its substance.

Informed critical judgments regarding the relationship between originals and copies, the greater or lesser authority of a given object or set of objects, and the work's meaning all become far more significant than the mere fact of accumulation. Following in the footsteps of the courtly patrons and collectors of the early modern period, new professional figures emerge alongside scholars by the end of the 19TH century, entrusted with guardianship over the remains of the past and armed with a battery of scientific and analytical techniques: archivists, museum curators, catalogers and librarians. The division between these figures and professional scholars is never absolute (as evidenced, for instance, by the role of attribution studies in art history and critical editions in the evolution of literary studies). But two parallel institutional worlds emerge that the digital revolution is reconnecting under transformed circumstances.

The reconnection assumes multiple forms: digital collection-building and curation on the part of individual scholars within and outside existing digital repositories as a form of scholarly practice; multimedia modes of argumentation that are object-based rather than discursive; conjugations of visible or audible digital media with physical objects in experiential exhibition spaces; the expanded publication of interpretive research results flanked by the archival documents and data sets that support them; large-scale collaborations that result in geospatially organized scholarly work; and critical editions of media artifacts that surround a primary object with multimedia objects rather than adopting only a text-based annotation system. It is nourished by the drive for research innovation, by the vastly expanded range of cultural materials now being produced and collected by institutions of memory, as well as by individuals and corporations; by the enhanced accessibility of these collections to both specialists and non-specialists alike thanks to their dissemination in digital form via the Internet; and by the crisis of print-based scholarly publishing and the potential for print-plus and post-print models that operate on scales unthinkable under the regime of print. *see* CASE STUDY 2 → 64 | 65; CASE STUDY 3 → 66 | 67; CASE STUDY 5 → 70 | 71

The Library of Alexandria is said to have held roughly half a million scrolls, representing works numbering in the tens of thousands. Twenty centuries later, Google Books has scanned, to date, around 14 million of the estimated 130 million printed books housed in physical libraries worldwide. What this means is that a contemporary scholar has at his or her fingertips access to 500 times the entire corpus of knowledge seemingly available in the ancient world without even consulting a second literary database or scanning the stacks of a major research library. The figure expands exponentially when we turn our attention away from "books" or "works" to other categories: print artifacts, letters, sound recordings, paintings, photographs, objects, telegrams, Web pages, email messages, blogs, tweets....

The scale and scope of these "collections," not to mention the accelerating creation of multimedia document collections in the present, so far exceeds the capabilities of traditional institutions of memory, not to mention the potential reach of scientific conservation methods, that enormous backlogs have become the norm. Despite sometimes heroic efforts to contain them, the amount of unprocessed or inaccessible materials in basements or off-site storage facilities are certain to grow within a setting in which information overload, the need to sift through and navigate vast data sets, and the proliferation of data trash are all givens. Critical curation is an essential scholarly practice in the print-plus and post-print world.

In common parlance, curation refers to the supervision and organization of preserved or exhibited physical items, although the term has origins in the theological domain, as in curates of the church who helped care for the souls of the dead. The term has exploded onto the contemporary scene, even invading business parlance: Slogans such as "ours is the age of curation" or "why calling yourself a curator is the new power move" are proclaimed in business blogs and reviews. What they are pointing to is the same urge animating the work of digital humanists: that the mere existence of vast quantities of data, artifacts, or products is no guarantee of impact or quality. To curate is to filter, organize, craft, and, ultimately, care for a story composed out of—even rescued from—the infinite array of potential tales, relics, and voices. In the Digital Humanities, curation refers to a wide range of practices of organizing and re-presenting the cultural record of humankind in order to create value, impact, and quality.

AUGMENTED EDITIONS AND FLUID TEXTUALITY

structured mark-up natural language processing relational rhetoric textual analysis variants and versions mutability

Critical editions—accurate and reliable versions of a text with an apparatus that presents and analyzes the evidence and source material to reconstruct and explain the original—have been a central part of the humanities for centuries. Some of the earliest critical editions were produced for the *Hebrew Bible* and the *New Testament* and aimed to become definitive editions for a wide readership. As the genre emerged and textual criticism matured, critical editions sought to foreground instabilities and uncertainties in the reconstructive process, providing ample room for commentary and annotation, the collation of textual variants, stemmatics, and studies of authorship, editing, transcribing, and translation practices.

Digital environments provide the ability to pull together many versions of a single work, tracking its development, noting its variants, and presenting the whole comparative array of witnesses. The basic tools for migrating texts into a digital environment are well-suited to such editorial tasks. The use of structured and/or tagged approaches to identify persons, themes, places, or features of a text provides a way to maximize the intellectual investigation of documents and to display these interpretations. As standards for mark-up (the tagging process used in transcriptions) have extended and improved, many nuances in textual analysis have become part of the set of interpretive elements. Not only can we identify what something is, but we can characterize its relation to other elements or entities (part of, derived from, a cousin of, a version of, and so forth). Under digital conditions, the very same procedures may now be extended to other categories of cultural objects, such as sound recordings, video, and film. *see* CASE STUDY 2 → 61 | 65

Fluid textuality refers to the mutability of texts in variants and versions whether these are produced through authorial changes, editing, transcription, translation, or print production. In a fundamental sense, then, texts have always been fluid and modular. But the advent of word processing drew intensified attention to this aspect of textuality. Writers were thrilled with the experiences of cutting and pasting whole portions of texts without retyping. The notion of transforming a work by changing its format and typographic font with the strokes of a few keyboard commands excited critical and creative imaginations. When it first appeared, hypertext was a foreign and intriguing concept, with nodes, links, and forking paths structured to create a multifaceted text in ways that had been tried in print formats but that took on an aura of novelty and promise in new media.

New dimensions of fluidity allowing for manipulation and machine processing of textual elements were introduced through natural language processing (NLP) and other tools for textual analysis. Global changes, searches, substitutions, counting, listing, reordering—these and other activities can be carried out through commands that treat a text as an object on which to perform operations that are somewhat at variance with conventional reading. In its most fluid state, a text file can be used to generate a nonverbal outcome. An ASCII string, or keyboarded text, can be turned into output in musical or visual form, or used to make a three-dimensional print, a pattern, or a design that serves as a template for another project in a medium far from that of verbal language. Texts are constantly flowed and reflowed, repurposed and reworked for different output streams and audiences.

With the increased fluidity of texts we see a corresponding change in authorial identity. We are witnessing a shift from the age of the individual voice to that of the collaborative, collective, and aggregated voice of the fluid text. Work in digital media frequently involves a composite warp and woof whose choral "textuality" reconnects the term with its origins in the world of textures and textiles.

With the emergence of the augmented edition, the fields of analysis and editing have grown and skills for creating scholarly editions are in increasing demand. The corpus of texts and other artifacts that comprise the record of human thought and culture is migrating from print and manuscript into digital forms, and, as that process advances, the need for editing and the opportunities for critical analysis continue to expand. Imagine that the surface of the screen becomes a deep space, and what appears at first to be a single page of a text or object extends through a multiplicity of embedded layers, each displaying a different facet of an argument or history of a work's production. Reader-viewers tier down and tier out, sifting through and engaging with, for example, a single word usage throughout a text, across a corpus, and across every book published in a given year. *see* CASE STUDY 2 → 64 | 65

The editing practices that extend into the augmented edition add other dimensions as well, allowing for a work to be understood within its larger field of cultural production; placed into the constellation of other productions and publications or artifacts of material culture; or situated within the documented events of an era. An augmented edition supports an array of arguments, with materials marshaled in demonstration of interpretations from a range of viewpoints or along a host of different lines of thought. The organization of argument in digital space creates new modes of content that are relational in their rhetoric. Sequence, juxtaposition, ordering, navigating, and analyzing are all features of the augmented edition. *see* CASE STUDY 1 → 62 | 63; CASE STUDY 3 → 66 | 67

SCALE: THE LAW OF LARGE NUMBERS

quantitative analysis text-mining machine reading digital cultural record algorithmic analysis

Although the Internet (meaning the technological infrastructure for transferring data over a distributed network of computers) is barely half a century old and the World Wide Web (meaning hyperlinked, hypertextual documents viewable in browsers) has only been around since the 1990s, it is striking to ponder the sheer volume of data they have "produced." Statistics vary. Some sources suggest there are more than 21 billion indexed Web pages, but the number of URLs indexed by Google is over one trillion; Google has scanned, cataloged, and made searchable more than 14 million books; Technorati has indexed well over a hundred million blog records since 2002, according to its annual report on the "state of the blogosphere"; JSTOR has over 7 million articles from more than 1,000 publishers; Facebook's repository is growing at a rate of 5 billion pieces of content per week, ranging from photographs and videos to news stories and blog posts; Twitter users produce, according to cofounder Biz Stone, one billion tweets per week. And these statistics do not even take into consideration the scope of other content produced and shared on the Web, such as email, not to mention content produced through participation in online community forums, chat groups, Instant Messaging, multiplayer gaming, and mixed reality environments such as Second Life. We are producing, sharing, consuming, and storing exponentially more cultural material—including texts, images, audio, and time-based data—than ever before. We are producing data at a rate that already outstrips our ability to store them and outpaces our ability to catalog, analyze, and archive these data in meaningful ways.

The humanities have historically been the province of close analysis of limited data sets: a literary study of a novel or poem, an art history monograph about a painter's *oeuvre,* an architectural critique of a Peruvian village's building styles. There have long been, of course, historical, generic, and stylistic studies with a broad scope. But for the most part there is a significant divide between the ways in which the humanities approach subject matter and the ways that more quantitatively inclined disciplines approach data. In the sciences, one of the key determinants is the law of large numbers, which states that the more times a researcher repeats a given experiment, the closer that researcher comes to determining an average value that defines the results of that experiment. Translated, this is simply a way of expressing confidence that ever-larger data sets will offer ever-more verifiable conclusions. Certain physical sciences deal with extremely large numbers such as atmospheric concentrations of carbon dioxide for climatology or genetic sequencing in biology. Until recently, such was rarely the case in the humanities.

When digital technologies allow for the storage and analysis of millions of books, billions of tweets, and hundreds of billions of interactions, the ways in which we can query and comprehend the cultural record explodes. Concepts, trends, actions, and the flow of human communication come into view at a macro scale. For instance, when huge numbers of related images can be parsed by machine, and the images themselves carry massive amounts of metadata, new questions can be asked about our relationships to the visual world. How do markets set the value of images? How do images in free circulation differ from those that carry price tags? Can we detect global patterns and regional differences in the ways that societies absorb and regenerate visual culture?

To answer such questions, we will have to design and employ new tools to thoughtfully and meaningfully sift through, analyze, visualize, map, and evaluate the deluge of data and cultural material that the digital age has unleashed: tasks that will require humanists to contend with text-mining tools, machine reading, and various kinds of algorithmic analyses. One way of navigating this process is through distant reading, a form of analysis that focuses on larger units and fewer elements in order to reveal patterns and interconnections through shapes, relations, models, and structures. It is a term that is specifically arrayed against the deep hermeneutics of extracting meaning from a text through ever-closer, microscopic readings. But, beyond distant reading, the time has come to entertain the possibility of machine reading, in which trends, correlations, and relationships are extracted through computational methods. Because information is being produced on a scale that far exceeds the faculties of human comprehension, it has become impossible to read, comprehend, and analyze the digital cultural record without the assistance of digital tools and methods. *see* CASE STUDY I → 62 | 63

To cite an example whose ethical stakes are high: What would it mean to subject the 52,000 Holocaust video testimonies in the USC Shoah Foundation Institute archives to machine reading and algorithmic analyses? Averaging two hours apiece, it would take a person 24 years to watch them all, assuming he or she watched 12 hours every day of the year. There is simply no way we can process and make sense of the volume of cultural data—including traditional printed materials—without the help of a computer to process, index, select, and cluster data on a comprehensible scale. But what are the implications of turning Holocaust testimonies into units of data, statistical analyses, and compact visualizations? Does this sort of quantitative analysis not inevitably, or perhaps by definition, subject the victims to further objectification, another dehumanizing process? Might there be an "ethics of the algorithm" that could mediate between the ethical demands of listening to individual Holocaust testimonies and the macrocosmic view enabled by a statistical representation of the total event? It is here that we need digital humanists to bring together the tools of technological analysis and the values, critical skills, and historical knowledge that animate the humanities disciplines.

DISTANT / CLOSE, MACRO / MICRO, SURFACE / DEPTH

large-scale patterns fine-grained analysis close reading distant reading differential geographies

Within the humanities, close reading has been a central practice that is premised on careful attention to features contained in a text, as well as its variations, history, transmission, possible meanings, and range of nuances. Close reading has its roots in the philological traditions of the humanities, but for more than a generation has often been equated with deep hermeneutics and exegesis, techniques in which interpretations are "excavated" from a text through ever-closer readings of textual evidence, references, word choices, semantics, and registers. The growth in size and accessibility of digital databases and concurrent advances in text-mining and what Lev Manovich has called cultural analytics have opened up new ways of creating meaning through distant reading. In the Digital Humanities, distant reading explicitly ignores the specific features of any individual text that close reading concentrates on in favor of gleaning larger trends and patterns from a corpus of texts. Distant reading is therefore not just a "digitization" or "quickener" of classic humanities methodologies. It is, rather, a new way of doing research wherein computational methods allow for novel sets of questions to be posed about the history of ideas, language use, cultural values and their dissemination, and the processes by which culture is made. Distant reading is almost not reading at all, but rather engages the abilities of natural language processing to extract the gist of a whole mass of texts and summarize them for a human reader in ways that allow researchers to detect large-scale trends, patterns, and relationships that are not discernable from a single text or detailed analysis. *see* CASE STUDY 2 → 64 | 65

Rather than pitting distant reading against close reading, what we are seeing is the emergence of new conjunctions between the macro and the micro, general surface trends and deep hermeneutic inquiry, the global view from above and the local view on the ground. The digital humanist is capable of "toggling" between views of the data, zooming in and out, searching for large-scale patterns and then focusing in on fine-grained analysis. While distant reading may be "new" insofar as computational techniques are involved in sifting through, organizing, and visualizing multitudes of data, it is worth remembering that humanities scholarship has long oscillated between and sometimes even conjoined these two approaches. After all, census data provide an overall statistical picture of demographics but tell nothing of the individuals who live in a given census tract; it is the task of oral histories, biographies, and psychological analyses to delve into the depths of the self. Similarly, a view from above in Google Earth allows a researcher to quickly pan over large regions of the Earth in order to discern surface structures of the built and natural environment, as well as overlay imagery and data sets such as National Oceanic and Atmospheric Administration satellite photographs of areas affected by storms, geographic information system (GIS) data relating to demographics, traffic trend data, and so forth. But these data become more meaningful when yoked with the stories of the people who actually live and have lived there, allowing researchers to not only "skim over" the surface of the Earth but also "drill down" into the micro-level temporal layers comprising the histories of every neighborhood, block, and home. Radically innovative approaches to mapping could emerge from within the Digital Humanities to create environments for exploring differential geographies and delving into heterogeneous geospatial representations, beyond simply registering the phenomenological aspects of space on conventional maps. It remains a challenge how to conceive, design, and implement such platforms. *see* CASE STUDY 1 → 62 | 63; CASE STUDY 2 → 64 | 65

CULTURAL ANALYTICS, AGGREGATION, AND DATA-MINING

parametrics   cultural mash-ups   computational processing   composite analysis   algorithm design

The field of cultural analytics has emerged over the past few years, utilizing the tools of high-end computational analysis and data visualization to dissect large-scale cultural data sets. Such data sets might include historical data that have been digitized, such as every shot in the films of Vertov or Eisenstein, the covers and content of every magazine published in the United States in the 20TH century, or the collected works of Milton, not to mention contemporary, real-time data flows such as tweets, SMS text messages, or search trends. Based on assumptions that meaning, argumentation, and interpretive work are not limited to the "insides" of texts or necessarily even require "close" readings, cultural analytics proposes that computational tools be used to enhance literary and historical scholarship. But creating models, visualizations, maps, and semantic webs of data that are simply too large to read or comprehend using unaided human faculties brings other questions. What parameters are used to incorporate cultural artifacts into data sets? Any conclusions based on these techniques are necessarily shaped, even determined, by these initial choices (e.g. if the gender categories for a census are only male and female, then how can we assess the percentage of transgendered populations?). Cultural analytics does not analyze cultural artifacts, but operates on the level of digital models of these materials in aggregate. Again, the point is not to pit "close" hermeneutic readings against "distant" data mappings, but rather to appreciate the synergistic possibilities and tensions that exist between a hyper-localized, deep analysis and a macrocosmic view.

Cultural analytics also broadens the canon of objects and cultural material under consideration by humanities scholars: Traditionally thought-of cultural objects are now digitized, marked-up, accessible, and shareable in multiple formats and on a variety of platforms, while "born digital" objects—whether tweets, blogs, videos, Web pages, music, maps, photographs, or hypermedia artifacts that combine many different media types—provide data for analysis and populate new forms of knowledge creation and curation. The "data" of cultural analytics are exponentially expanding in terms of volume, data type, production and reception platform, and analytic strategy, making it all-the-more important that humanists are engaged with the design of algorithms, mining and visualization tools, and archiving techniques that foreground questions of value, interpretation, and meaning.

Aggregation of large-scale amounts of information allows data or files to be merged and then outputted into displays that highlight distinctive features such as data points, clusters, and trends. Structured data lend themselves to this processing, since one might easily take dates, places, quantitative information, names, or other elements from a set of files and create an analysis of its contents. Tracking network traffic, or money flows, or resource depletion, or economic trends works well in aggregate. In text processing, looking at word frequency and use (the n-gram approach) is a way of aggregating information and data. The aggregate subsumes individual instances, extracting information from the whole. Cultural mash-ups often aggregate materials in novel ways that allow digital manipulation to repurpose the sources.

Composite analysis preserves individual elements but uses the patterns among them to show something about the whole set of discrete elements. The information and data remain linked to the individual instance rather than being extracted from it into a larger whole. The affordances of large-scale displays, in which thousands of individual images or artifacts can be shown and accessed creates a composite environment. Use of computational methods to discern patterns among such large corpora is essential, though figuring out what the particular purposes or research questions are that can be answered by such techniques necessarily depends upon recognizing that analysis and processing follow from the fundamental decisions about what constitutes the data and the ways in which these data are structured.

Finally, data-mining is a term that covers a host of techniques for analyzing digital material by "parameterizing" some feature of information and extracting it. This means that any element of a file or collection of files that can be given explicit specifications, or parameters, can be extracted from those files for analysis. The "mining" of these data often depends on creating a display of the results as statistics, texts, or in an information graphic known as a data visualization. Understanding the rhetoric of graphics is another essential skill, therefore, in working at a scale where individual objects are lost in the mass of processed information and data. To date, much humanities data-mining has merely involved counting. Much more sophisticated statistical methods and use of probability will be needed for humanists to absorb the lessons of the social sciences into their methods. *see* CASE STUDY 2 → 64 | 65

VISUALIZATION AND DATA DESIGN

data visualization mapping information design simulation environments spatial argument modeling knowledge visual interpretation

In recent years humanists have become increasingly involved in what is often referred to as the "visual turn" in scholarship, sometimes correlated with the "spatial turn" that has favored mapping. As digital tools have become prevalent, the interest in "reading" the visual has extended to "authoring" the visual—using visual means to express intellectual concepts. What might it mean to make a visual argument, for instance, or to shape a concept through graphical means?

Currently, visualization in the humanities uses techniques drawn largely from the social sciences, business applications, and the natural sciences, all of which require self-conscious criticality in their adoption. Such visual displays, including graphs and charts, may present themselves as objective or even unmediated views of reality, rather than as rhetorical constructs. Much could be learned from the visual languages and semiotic critiques of art, architecture, and design. Visual special effects, which add more to spectacle than to legibility, are suspect, and information graphics conceived without some professional competence in their design are often unintentionally misleading. By the same token, visualizations designed to specifically address the communication needs of humanities research will only be created if humanists become actively engaged in their design.

The visualization identified here refers primarily to graphical or rendered visual interpretations rather than photographs or films, which have their own unique history and rhetorical qualities. Visualization is intellectually distinct from illustration, the employment of a graphical feature, photograph, map, or other representational device to elucidate, explain, or show something in a text. In the latter case, the text still assumes priority, and the illustration is meant to summarize an argument, provide a reference point, or corroborate the text. While visualizations may illustrate data through processes of aggregation and distillation, visualization in the Digital Humanities takes several different forms, all of which are arguments in themselves and must be evaluated in terms of the rhetorics of information design and display.

The use of graphs, charts, diagrams, and other visualization techniques is often associated with data visualization, the expression of quantifiable or quantitative information in graphic form. But the models of statistical expression, such as bar and pie charts, came from the world of 18TH century "political arithmetic" and provided a rich and much developed legacy that extended the vocabulary of much older visual forms of diagrams, grids, and trees. Business, governments, and administrative organizations all made use of these forms to express quantitative analyses in legible formats. Informational and statistical visualizations engender the rhetoric of clarity, precision, and fact, though they are, of course, constructed interpretations. When done well, they can make persuasive visual arguments, allow something new to emerge, or even be subverted for poetic effect.

Visualizations of data that are produced computationally tend to be derived from large-scale data sets such as social networks, digitized corpora, and demographic data. The visualizations, either custom-built (e.g. network analysis diagrams) or created for use in an online environment (e.g. Many Eyes), may be used as analytical and interpretive tools—to reveal patterns or anomalies or concurrences—or they may be produced to illustrate findings or serve as the distillation of an argument. Of course, the structures of the data and the questions that are asked of them will, inevitably, determine the visualization produced and the answers obtained. Perhaps it is of little surprise, then, that data visualizations tend to take the established forms mentioned earlier—charts, diagrams, grids, or trees—although we are increasingly immersed in a world of graphical possibilities that have yet to be realized. *see* CASE STUDY 3 → 66 | 67

Mapping is a distinct form of visualization built on the history of cartography; ideologies of discovery, ownership, and control; levels of abstraction; scale; relations between the real and representation; symbology; visual signposting; perspective;

and coordinate systems. Mapping in the humanities ranges from historical mapping of "time-layers" to memory maps, conceptual mapping, community-based maps, and forms of counter-mapping that attempt to un-ontologize cartography and imagine new worlds. In the 1950s, the members of the experimental Situationist group developed an approach to experiencing urban spaces that they termed "psychogeography." The immersive and experiential wandering advocated by the approach gave rise to a handful of maps that suggested flows and movement through space as a ludic, exploratory exercise that could result in a new critical awareness of urban environments. Similarly speculative, cognitive maps are used to model experience in many domains of human life where qualitative properties are given dimension and formal value in visual form without any need to ground them in quantitative information. *see* CASE STUDY 1 → 62 | 63

A human life may have many such experiential dimensions in which affective properties shape the intellectual argument and give rise to a graphic form that shows the size, scale, proportions, orientation, direction, or figurative shape of knowledge unfolding over time. Maps, animations, and visual images from the vast inventory of human imaginings have much to offer to contemporary scholars re-imagining their own concepts of intellectual argument. The pictorial conventions of visual representation may well be repurposed, just as the organization of cabinets of curiosity, antiquated libraries, personal spaces of study, commonplace books, or other instruments of memory, argument, and rhetoric are finding their place again within the broader understanding of how we produce and represent intellectual arguments and model knowledge.

Experiential visualization uses movement through the time and space of a three-dimensional world as the primary mode of engagement. Historical simulation environments can take a viewer into an immersive environment (or, at least a virtual one), creating the experience of walking through, for example, a Chinese farming village during the Han dynasty or the Roman Forum in late antiquity. Historical simulation environments don't represent the past "as it really was"; instead, they foreground interpretation, analysis, and experimentation, allowing new research questions to be asked and hypotheses to be tested using a wide range of variables. For instance, one may employ time-sliders to visualize when and where certain buildings came into existence or to investigate kinetic aspects of events in time-space environments, such as parades, funeral processions, orations, and protests. Experiential visualization is not a simple mimesis or positivistic reconstruction of a historical reality, nor is it a simple augmentation of a real-world site, but rather an investigation of a state of knowledge. *see* CASE STUDY 4 → 68 | 69

Visualization can be used in many other ways to sketch out an argument or to map its constituent parts or even to model its initial formulation. Visualization has the power to unleash imaginative and conceptual potential. By identifying elements of a system and thinking about how they relate to each other sequentially, or hierarchically, or relationally, humanists discover ways of modeling knowledge that were not part of their textual training. As with so many aspects of digital work, the strengths of these techniques are amplified when they are in dialogue with, rather than opposed to or exclusive of, traditional methods. The use of visualization or distant reading, for example, in concert with attention to individual texts, or aggregation techniques in dialogue with studies of outliers and anomalies can provide valuable contributions to the discussion of meaning-production that could not be obtained using only one or the other method. Knowing what to read and visualize as well as how to read and visualize forms is at the basis of digital literacy and the assessment of meaning in these new formats. *see* CASE STUDY 5 → 70 | 71

LOCATIVE INVESTIGATION AND THICK MAPPING

spatial humanities digital cultural mapping interconnected sites experiential navigation geographic information systems (GIS) stacked data

Traditional scholarship in the humanities moved among a few select sites for research and teaching: the library, the archive, and the classroom. The "holdings" of the library were just that: holdings held for the initiated who had the privilege of access and use. Scholars made pilgrimages to special collections to view artifacts or read rare books, often examining these objects under guarded conditions that were established to limit access and thereby preserve the safety—and aura—of the original. The seminar room or the lecture hall was the primary site for the transmission of knowledge mostly in a single direction: professors professed knowledge and students consumed it.

Today, the boundaries of the library, the archive, and the classroom have become more porous, interconnected, and globally extensible. Countless new sites for knowledge creation and dissemination have emerged, bringing scholarship into communities and communities into the academy. Libraries have allowed millions of their volumes to be digitized and have opened up their collections via Web services, making them available to the digitally enfranchised public. Some archives are following suit, removing physical and virtual walls that once restricted use. The traditional learning space of the classroom has been rethought in ways that promote interactivity, discovery, and co-creation, often through real-time feedback mediated by social technologies that blend physical and virtual spaces. Courses have been taught, for example, in virtual worlds with avatars participating from around the globe, connecting the physical space of the classroom with the infinitely expansive and fluid realms of cyberspace. Webcast or web-linked teaching is now commonplace.

But a few caveats are in order. The networked world is a patchwork, very much marked by social and economic inequalities; access and participation are hardly open to everyone. Different zones are governed by distinct attitudes toward cultural property, licensing, and pressures of sustainability. Cost-recovery models exist even in the most elite sectors while many individuals and communities have limited connectivity. Just as in the realm of bricks-and-mortar education, inequities abound. Diverse and competing interests will continue the struggle for control.

As the contours of scholarship are undergoing a fundamental remapping through collaborations in which researchers can curate, narrate, annotate, and augment physical landscapes, the boundaries of inside and outside have become fluid. The interior realm where curators make arguments in space through the meticulous staging of physical objects, supported by labels, wall text, and installation architectures can now be enriched by media that draw the outside world into the gallery. Likewise, visitor itineraries can now be extended out into the surrounding landscape in ways that apply traditional curatorial skills to the shaping of paths through the physical world. Data landscapes can be curated in the physical space of a city, allowing a user with a GPS-enabled mobile device, for example, to listen to geo-coordinated soundscapes curated by musicians while walking down a sidewalk or to follow in the footsteps of the dead and hear stories told by generations of immigrants about a neighborhood. Such locative investigations bring together the analytical tools of geographic information systems (GIS), the structuring and querying capacities of geo-temporal databases, and the delivery interfaces on GPS-enabled mobile devices. *see* **CASE STUDY 5** → 70 | 71

This attention to place has resulted in the emergence of a significant sub-field of the Digital Humanities variously called "Digital Cultural Mapping" or "Spatial Humanities." It is here that geographic analysis, digital mapping platforms, and interpretive historical practices come together to form richly textured, multidimensional investigations of place. Unlike conventional approaches to mapping, which tend to be positivistic and mimetic, these practices of thick mapping in the Digital Humanities place a primacy on experiential navigation, epistemologies of representation, and the rhetorics of visualization. After all, a map is a visualization or representation of a group of relations (and structuring assumptions) that present a state of knowledge. The map may or may not have a referent in the "real world," but it does make an argument, and in the digital realm it becomes an interactive site for creating, representing, and navigating knowledge. Digital maps are essentially navigable layers of spatial data rendered visually, ranging, for example, from demographic and census data to location-specific video histories, Twitter streams, and historical map layers.

Such maps are not meant to be static representations or accurate reflections of a physical reality; instead, they function as stacked representations in which one representation is linked or keyed to another. Within a dynamic, ever-changing environment, new data sets can be overlaid, new annotations can be added, new relationships among maps can be discovered, and, perhaps most importantly, missing voices can be returned to specific locations through "writerly" projects of memory that the participatory architecture of Web 2.0 applications has made possible. Thick mapping thus enables an unbounded multiplicity of participatory modes of storytelling and counter-mapping in which users create and delve into cumulative layers of site-specific meaning. Far from the Apollonian eye looking down from a transcendental view, thick maps betray the contingency of looking, the groundedness of any perspective, and the embodied relationality inherent to any locative investigation.
see **CASE STUDY I** → 62 | 63

THE ANIMATED ARCHIVE

user communities   permeable walls   active engagement   bottom-up curation   multiplied access   participatory content creation

Derived from ancient Greek ἀρχεῖον (*government*) and the late Latin word *archivum*, the English derivative *archive* now refers not just to public administrative records but also to the entire corpus of material remains that the past, whether distant or close, has bequeathed to the present: artifacts, writings, books, works of art, personal documents, and the like. Its semantic field also encompasses the institutions that house and preserve such remains. In all of these meanings, archive connotes a past that has severed its ties with the present and has entered the crypt of history only to resurface under conditions of restricted access.

The Digital Humanities offers new challenges and possibilities for institutions of memory such as archives, libraries, and museums: process-based concepts of "living" archives of the present; approaches to conservation and preservation based upon multiplying (rather than restricting) access to the remains of the past; participatory models of content production, research, and curatorship bringing together professional and citizen scholars in team-based projects that interpret the cultural patrimony as a public good; augmented approaches to programming and informal education that promise to expand traditional library and museum audiences and bring scholarship into public view; and enhanced means for vivifying and promoting active or experientially augmented modes of engagement with both the past and the present. Of course, the past was never really past; it always already belonged to the present. And digital toolkits and the expanded compass of humanistic scholarship provide some distinctive avenues for investing the present's stewardship of the past with the attributes of life. They hold out the promise of animating the archive.

Accumulation is no longer enough to ensure the survival of the cultural patrimony. Objects that sit in storage, though they may have a potential afterlife, disappear into the ever-expanding heap of cultural remains, entering a limbo that in no essential way differs from being lost. So the "animation" of archives stands for a series of strategies for launching that afterlife from the very moment of archival processing. This implies a user-centered approach to the construction of archives that builds a multiplicity of use-scenarios into the very architecture of the archive; breaks down partitions between collections and bricks-and-mortar institutions (through, for example, open application programming interfaces); engages real or potential user communities from the outset (in processing, tagging, and metadata development); and integrates curatorial and content-production tools into access portals. *see* CASE STUDY 3 → 66 | 67

Embedded within the constellation of possibilities just evoked is a sort of Copernican revolution with respect to the roles performed by libraries and museums in the modern era. New conjugations of inside and outside, scholar and citizen, curator and viewer are emerging, with social technologies challenging conventional ideas of ownership, restricted use, storage and display, content creation, and curatorial control. With the shift in focus from data retrieval—essentially "top-down billboarding"—to bottom-up working and reworking of content, whether in the form of texts, still or moving images, audio, or other media, every library and museum becomes adjacent to a public square as big or as small as they choose. It also marks the beginning of an inversion which some will welcome and others will decry. Whereas the virtual was once subordinated to and cast in a supporting role with respect to the physical, we are now seeing new couplings in which an institution's virtual footprint may exceed its physical edifice and the community that it serves may be worldwide, overlapping only in small part with potential or actual physical visitor/user populations. This is one of the great opportunity spaces that the Digital Humanities opens up, giving archivists, librarians, and curators a chance to not simply enlarge but completely re-envision their communities, publics, and missions. Every public institution has already been transformed into a glocal enterprise, local and global at the same time. Glocalization will only accelerate over the coming decades. *see* CASE STUDY 4 → 68 | 69

In sum, the memory palaces of the 21ST century will have much more permeable walls than their 19TH and 20TH century predecessors. This is also to say that they will be much bigger both from the standpoint of the physical territory that they cover and the corpora of information that they harbor. For example, the Digital Humanities harnesses the expressive power of worlds like Google Earth and three-dimensional virtual environments, and deploys the ever-increasing availability of wireless bandwidth to interact with ubiquitous computing devices equipped with GPS technologies that can calculate and annotate embodied, physical locations within inches. This is the future of knowledge, where culture and social and political practice will emphasize embeddings of the virtual within the real, where actual physical landscapes will be curated just as if they were an art gallery, and where we will be surrounded and enveloped by the collaborative and distributed building of annotations on, and overlays of, the physical world. This is a future that is already with us. The challenge for scholarship and institutions? To build platforms and collections out into these and other domains of intersection between the virtual and the physical in ways that reinforce not only access and outreach but also establish new models of imagination, quality, and rigor.

DISTRIBUTED KNOWLEDGE PRODUCTION AND PERFORMATIVE ACCESS

global networks ambient data

collaborative authorship interdisciplinary teams use as perfomance crowd-sourcing

The myth of the humanities as the terrain of the solitary genius, laboring alone on a creative work, which, when completed, would be remarkable for its singularity—a philosophical text, a definitive historical study, a paradigm-shifting work of literary criticism—is, of course, a myth. Genius does exist, but knowledge has always been produced and accessed in ways that are fundamentally distributed, although today this is true more than ever. It is not uncommon for dozens of people to work on a Digital Humanities project, each contributing domain-specific expertise that enables a research question to be conceptualized, answered, and then re-conceptualized and re-answered. A team of database developers and data management experts may come from a school of information sciences, while interface designers may come from the arts, content developers may come from history departments, and coders may come from the computational sciences. Each member of the team works with the technical lead and project director who collaboratively articulate the technical and functional specifications for the project. In the end, when a project is deployed, there may be dozens of "authors"—ranging from professors and librarians to student programmers, interns, staff, and community members—who contributed to its development. Some long-term projects are the work of generations of students and scholars. Distributed knowledge production means that a single person cannot possibly conceive of and carry out all facets of a project.

Analogously, distributed access means that the audience for the project can engage with its content via multiple access points and platforms. In fact, every engagement is a performative instantiation of knowledge. With the surge of mobile devices and distributed computing, ambient networks present new possibilities for accessing information and interacting with knowledge. While we access most digital information on a screen, the means by which information circulates to find its connection to those screens is distributed across wired and wireless networks, with data shuttled seamlessly between the cloud and our local machines. Though our perception of them is limited to display devices, data streams fill the air. Their presence in and among the many other features of the physical environment makes their integration into lived experience a possibility. Human and cultural knowledge will interpenetrate the natural and built environments with increasing degrees of saturation. Access to interpretive materials, cultural history, geographical and geological knowledge, historical dimensions, narrative facts, biographical information, and the stories of events lived and experienced in our shared spaces will be a way to enhance the engagement with the real. Or, in other instances, they may provide solace, consolation, companionship, and fellowship through communicative exchange. What is certain is that knowledge production, access, and dissemination are becoming ever-more distributed processes across high-speed, mobile networks that operate seamlessly at all levels. *see* CASE STUDY 4 → 68 | 69; CASE STUDY 5 → 70 | 71

When knowledge exists in iterative form across global networks and local access points, with many versions and expressions of cultural information taking shape in a process whose life cycle is ongoing, then any access to that knowledge is a performance, an instantiation. Just as any reading of a book or a script or any viewing of a film or any playing of a score is a performance of that work, the same is true of digital works. In fact, every use of a file is different; no two files are ever the same, and the very act of opening and displaying a file is a performance of a work,

→ HUMANITIES GAMING

a unique instantiation in historical and social space. The difference between the performative reading of a work in an analog world and in a digital one registers dramatically how the lines between reading and authoring blur. When the material substrate records the performative variations of each instantiation, then the act of reading or viewing contributes materially, not just virtually, to the work. When and how such traces will be recorded has yet to be seen and will constitute fertile ground for research and publishing in the Digital Humanities, but the possibilities for crowd-sourced engagement with editing, proofreading, translation, and critical assessment are bringing this process into view. *see* CASE STUDY 3 → 66 | 67

HUMANITIES GAMING

user engagement rule-based play rich interaction virtual learning environments immersion and simulation narrative complexity

Imagine being on the streets of a South African township as it explodes in violence after the apartheid-era government switched the language of education from English to Afrikaans. This is the experience Hamilton College students have when they play the immersive game Soweto '76, one that deepens empathy and enlivens class discussions of race, power, and education. At Dartmouth, students compete furiously against each other to tag the materials they find in online archives. When these students play Metadata Games, they are encountering an open-source project that uses the affordances of gaming to build more robust archival data systems. King's College London students create avatars in Second Life and then reconstruct historical stages from the classical Roman Theater of Pompey to Shakespeare's Globe Theatre in London. At Duke and other participating universities, students play "Virtual Peace," a collaborative simulation game in which players analyze complex situations posed by international crises in order to learn how to make effective decisions. Digital Humanities gaming has begun to successfully engage with historical simulation, virtuous cycles of competition, and the virtual construction of learning environments.

But games of any type have never been held in high repute by academia. Relegated either to athletic departments as mere sports, or to the realm of leisure time as diversions, they have only recently begun to be taken seriously as both an object of study and a career for which to be trained. Gaming demonstrates a capacity that could transform Digital Humanities pedagogy. This is due to many factors, but two in particular stand out. The first is the explosion of processing power and connectivity. Not only are game-world simulations compelling visually and interactively, they are also capable of functioning in real time with multiple participants spread around the world. This braiding of capacity and reach, made possible by ever-increasing processing speeds, the ubiquity of networks, and mobile connectivity, yields highly engaging forms of immersion and simulation. This rich interaction can be yoked to any content, from the expected adrenaline thrills of first-person shooters like Halo to detailed alternate reality games (ARGs) like World of Warcraft. Given the fungibility of content and the consistency of user engagement with well-designed games, "humane" and "serious" games are likely to keep pace with technological advancement. The second fact to consider here is the acculturation of a generation of students who have literally grown up gaming. They value interactive programs that engage their attention while at the same time deepening their understanding of meaningful subject matter.

But what exactly do such students expect and what constitutes a successful educational game? Games are rule-based. They offer copious feedback. They are essentially voluntary, running on enthusiasm and begging for engagement. Games are also quintessential delayers of gratification: Give players the freedom to achieve their goals in the quickest, most rational way and satisfying game play withers. It is the obstacles overcome and the levels mastered, the reward for tasks accomplished and the rules obeyed which constitute the satisfaction of play.

Recent developments in new-media studies and narratology have removed some of the stigma that was once attached to gaming within the academy, but digital games are still considered by many in the humanities as frivolous (and monstrously violent to boot). It becomes harder to maintain this perspective as the narrative complexity, play strategy, and game "feel" (as developers call the gestalt of gamer-and-game interaction) become more developed, culturally significant, and even world-enriching. As we have seen, games in the Digital Humanities already exist that are exploring interactive models of learning and ways of critically grappling with the human experience. The challenge for the future is to take the gamesmanship of humanities research—its pursuits and pleasures, competitive drive, and seductive engagements—as the basis for games of scholarship. *see* CASE STUDY 4 → 68 | 69

CODE, SOFTWARE, AND PLATFORM STUDIES

narrative structures code as text computational processes software in a cultural context encoding practices

Code studies, along with the related study of software and platforms, bring humanistic close-reading practices into dialogue with computational methods. The operations of computational media are created through the interaction of hardware and software. These work according to protocols structured into their organization as code. The study of code is driven by an interest in exposing the ways constraints make certain things possible, and exclude others. But is code a text? If so, what kind of text? Should we assess the aesthetic properties of code the same way we discern the value of any other artistic composition? Or should we condemn code work as mere craft or technique? Debates are heated, with passionate partisans on all sides. The alphanumeric system is already a code, so the heralding of a "new" field of code studies may seem inflated.

Code and software scholars begin their study with the history of encoding practices, in particular those methods that make an operation happen, such as the punch cards used to set the patterns of weaving on Jacquard looms or the programs in early computers enabled by stacks of cards whose punched openings allowed circuits to be blocked or completed. The basic binary language of digital media is the foundation of all programming code, but software and computer languages have their own history as forms with grammar and syntax. The study of software traces developments from switch settings on mid-20TH century mainframe computers to the creation of the assembly and compiling languages that underpin many of the scripting languages and much of the object-oriented software written today. The layers of software between the operations of a machine and the instructions given it by an operator offer a fascinating archaeological study, with cultural conventions often holding as much weight as technological advantage.

Scholars fascinated by the encoded protocols and instructions that constitute the language of software also look at the cultural contexts in which business, defense, or communications industries fueled the development of increasingly sophisticated approaches to encoding. The algorithm, a set of step-by-step instructions, is the heart of software programs, but these instructions have to be translated into a binary language that the computing hardware understands. The organization of processing units, the workflow cycle through circuits and transistors, the use of active buffers and parallel processors—all of these pieces of hardware interact with software in particular ways that have affordances and hindrances that vary from platform to platform. Critical approaches allow understanding of these elements as objects of study, almost as if one were reading them as text.

A particular fascination with game engines and their narrative structures fuels one area of code studies. The analysis of narrative and multi-player activities in a complex set of chained and interdependent interactions requires chunking of game elements at critical nodes or decision points. The ways this is achieved is itself a complex process—a game of sorts—in which the skills of narratology meet the worlds of probability and possibility in a combinatoric universe that must move seamlessly from one moment of illusion to another. The way this is engineered and designed elicits a fascination akin to that of expertly constructed aesthetic artifacts in any other era, such as novels or plays. Likewise, the engagement of a substantial literary community with the poetics of code has created a body of critical work that addresses the aesthetics of programming in its own right. All of these approaches offer analytic engagement with computational processes as forms of composition, exposing their complexity through careful reading, construction, and attention to structure. *see* **CASE STUDY 5** → 70 | 71

DATABASE DOCUMENTARIES

variable experience   user-activated   multimedia prose   modular and combinatoric   multilinear

Digital Humanities genres include multimedia critical editions; interpretive work with expanded data sets published alongside their interpretive outcomes; conjugations of the digital and the physical, the desktop and the streets; and expanded definitions of knowledge that exercise not just sight, but the entire human sensorium. Within this set of emerging composite forms, the database documentary occupies a central position. It is a genre that has continued to evolve in dialogue with shifts in the technology of interactive media.

Cinematic documentaries work with image and sound materials that, however mediated or massaged, claim an indexical relation to the world. That is to say that they work with "real-world" materials captured, filtered, and threaded into a linear narrative artifact in the medium of film. In order to craft such a linear narrative, large amounts of footage must be shot as part of the research and development process. By necessity, most of this footage must be thrown out or reduced to a few choice sequences, given that a small core of materials must make up the story's backbone. Only one story can be told well, even when the intended single "story" turns out to be a densely wound skein of stories, each overlapping with the next.

The database documentary also works with materials of documentary value, but on an expanded scale. Database documentaries are modular and combinatoric, branching and hypertextual, often structured more like a multimedia prose piece than a film. Consisting of a series of tracks through an actual or virtual database, the documentary can be built out of a wide range of media types: not just film and video, but also sound, static image, text, animation, actual documents (or their digital equivalents), even live or dynamic feeds from the World Wide Web. Database documentaries are multilinear. They are not watched, but rather performed by a reader/viewer who is provided with a series of guided paths; and, unlike the cinematic documentary, which is free-standing, database documentaries may be built on multiple, overlapping databases. Or they may even consist only of pointers that send out calls, through open APIs, that retrieve materials hosted externally. The paths are reversible, allowing for trackbacks to the sources from which individual documents are drawn and/or to external resources. Inclusions as well as exclusions can be exposed to view, thereby creating an experience that is dynamic, active, and user-centered. Temporal sequence, duration, and sound levels, as well as the presence or absence of elements of the critical apparatus are firmly in the reader/viewer's control. *see* CASE STUDY 4 → 68 | 69

The multilinear character of database documentaries creates a different series of opportunities and challenges with respect to cinematic documentaries. Given that multiple intersecting story lines are present in database documentaries and that they are user-activated, a far greater fluidity of movement and pacing must always be presumed, much as in the case of visitation paths through physical exhibits. Conferring unity upon such a variable experience can be difficult, as can the building of cohesive story lines. This said, the possibility of marshaling crisscrossing sets of data to tell interrelated stories offers powerful new modalities of scholarly argument as well as imaginative expression. The database documentary remains one of the most venerable of new media forms, with early expressions such as the Interactive Cinema Group at MIT in the late 1980s, the Labyrinth project at the University of Southern California, and the pioneering work of United Kingdom-based Blast Theory group in the field of so-called "live" documentaries.

REPURPOSABLE CONTENT AND REMIX CULTURE

participatory Web   read/write/rewrite   platform migration   sampling and collage   meta-medium   inter-textuality

The ease with which content can be repurposed in digital form extends the capacities of the medium to function as a meta-medium. Photography has that property, with its ability to record and reproduce drawing, painting, printmaking, and other visual formats. Now, the digital environment serves as the simulation machine that is able to re-create and imitate other formats. But it also allows content to be migrated from platform to platform, to be used in a variety of outputs and for a range of readers and forums. Figuring out how to write texts in a modular manner that will allow them to be recombined for different levels of interest and readership as well as different degrees of detail and granularities of argument (not to mention output and display device) is still a challenge—and represents another fertile field for the Digital Humanities to explore. The realization that print on-demand and online access are complementary modes to traditional print rather than competitive ones is already well-recognized by the publishing community as well as readers and authors. Artists also engage opportunistically with the possibilities of different venues and formats, so that their range of expression might include gallery works that are unique, printed versions for larger distribution, and online exhibits of the same work to reach yet another audience. The work is a distributed effect of each and all of these aspects rather than being limited to any single part of this continuum. *see* CASE STUDY 5 → 70 | 71

Remix culture is a hallmark of the participatory, programmable Web in which a "read-only" ethos has been surpassed by one of "read/write/rewrite." In much the same way that early textual scholarship used citations and annotations to extend authority on copied manuscripts, remix culture uses digital sampling and collage techniques to create derivative original works with a complex trail of associations, inter-textual references, and critical trajectories. Authorship is multiplicative and dissemination happens across the Web as others add to, borrow, remix, and republish the work. Best known in music, remix culture extends to photography, film, graphic design, software development, data curation, and many other realms. In essence, with the tools of both production and consumption in the hands of the public, an ever-expanding space of design and curation allows bits of data and intellectual property to move and be remixed in creative ways. *see* CASE STUDY 4 → 68 | 69

The university, however, still places a primacy on the singular nature of originality of scholarship and on clear lines of demarcation for authorship. In fact, the institutional structures for generating, evaluating, and legitimizing knowledge have barely embraced repurposable and remixable intellectual culture. Perhaps this is because the institutional frameworks in which this knowledge is produced and evaluated have hardly changed over the past century. What if departments could be remixed as easily as digital music samples? What if curricula had life cycles like software? What would an open-source humanities division look like? For one thing, disciplines, departments, and administrative structures would receive date stamps and would need to innovate in order to survive.

The objections are, of course, easy to mount: Without the long-term stability of a department, how can we prepare students for a field? How can we be sure that they have learned "the content" of a discipline, and how can we possibly credential students with degrees if they are participating in departments that will no longer exist in a matter of years? These objections, we believe, are based on assumptions that have traditionally valued "the what" (a determinate and relatively static set of knowledge objects or canon of artifacts) over "the how" (a flexible—even nimble—mode of thinking that privileges design, experimentation, risk-taking, and creative problem-solving). This is not to say that knowledge in a field is irrelevant, for the contrary is true. It is to say, however, that universities will serve their students best by credentialing the skills necessary to creatively conceptualize and solve problems: a knowledge base grounded in making and experimentation, and a social disposition that fosters collaboration with diverse partners. It is here that the core values of the humanities and the generative potential of the digital come together in the poiesis of world-making.

PERVASIVE INFRASTRUCTURE

extensible frameworks   heterogeneous data streams   polymorphous browsing   cloud computing

With the emergence of standards-compliant Web services and dynamic cloud computing, massive data sets can be shared and accessed across networks. Web services are essentially machine-to-machine communications that allow various types of data to be accessed through specific queries. For example, a map Web service might allow a user to access census data or historical maps stored on one server from any computer able to send the appropriate query to the service; users will not receive the "actual" maps or entire copies of the data but rather access to the maps and data through calls to the service provider. The data can, then, be rendered and viewed in various interfaces, such as on a Web page, in a geo-browser, or in another visualization application. Cloud computing provides an (almost infinitely) extensible framework for massive data storage, access, and retrieval from any computer connected to the network. The metaphor of the cloud signifies the seemingly ethereal data that can be pushed and pulled through the sky, but in reality it translates into mega data centers, storage systems, and networked Web architecture to facilitate data exchange.

What does this mean for the Digital Humanities? Foremost, it means that it is now possible to share the entire data sets of research with the scholarly community and the public-at-large. In disciplines such as anthropology, archaeology, and classics, researchers may produce millions of discrete data points over the course of a project, ranging from survey and excavation data to fieldwork documentation through integrated geographic information systems. Rather than summarizing the results of a project and drawing conclusions, researchers can make the entire data set available online, enabling other users to test hypotheses and even to add to and edit the "original" data set and accompanying metadata. Openness has benefits, but caveats about validity of data, privacy, misappropriation, and other ethical concerns are also in order. Secondly, through polymorphous browsing, users can access, manipulate, and analyze massively heterogeneous data streams, following trails of association that lead out and go deep. What this means in practice is that search and discovery tools are able to identify, aggregate, and integrate data from completely disparate sources across archives, libraries, and repositories and present these data in ways that are customizable for the needs of a given researcher. One can expand and contract, tier out and drill down through a portal that can access the world's information regardless of where it resides. For scholars of literature, for example, it means having access to every word in every edition of every book ever published and customizing a search to answer a research question that, recursively, becomes part of the data of the system itself. *see* CASE STUDY 3 → 66 | 67; CASE STUDY 4 → 68 | 69

UBIQUITOUS SCHOLARSHIP

augmented reality   web of things   pervasive surveillance and tracking   ubiquitous computing   deterritorialization of humanistic practice

As these emergent genres and methods illuminate, the forms that knowledge assumes can no longer be considered givens. The tools of humanistic inquiry have become as much objects of research and experimentation as have the modes of production and dissemination of knowledge. Statistical methods press against one edge of the qualitative human sciences; graphic and information design press up against another. Real time, massively participatory role-playing games create another force field exerting influence from the arts and gaming worlds. Laboratories arise with a collaborative, team-based ethos, embracing a triangulation of arts practice, critique, and outreach as they merge research, pedagogy, publication, and generative practices. The once-firm boundary lines among libraries, museums, archives, and classrooms have become increasingly porous as scholarship, no longer limited to print and the lecture hall, has started to shuttle back-and-forth between the stacks and the streets.

Location-aware smartphones and other mobile devices have a key role to play in this deterritorialization of humanistic practice. Thanks to their ubiquity, it has become possible to couple Web-based knowledge resources to physical locations in ways that would have been hard to imagine only a decade ago. This means that scholarship in fields such as history, urban studies, architecture, art, design, and literature can now curate, narrate, annotate, and augment the physical landscape with a multitude of Web-based archival sources. Such scholarship speaks to multiple audiences and leads multiple lives. A website may be remixed as an electronic publication for use on location-aware mobile devices and later become a print artifact; the website and mobile "edition" can be built for further curation and extension on the part of end-users who can embed datascapes anywhere, at anytime. Augmented reality applications allow mobile devices to combine geolocation information and enhanced imagery in a layered, site-specific presentation of events and interpretations. Imagine a time-machine application that shows your neighborhood in a fast-forward sequence from Jurassic times to the present; or think of sensors in a natural environment that expose the geological and industrial processes that formed what is before your eyes; or consider simultaneous and automatic translation applications that remove linguistic barriers to signage and information in a foreign script; or imagine the "web of things," in which every physical entity—from the book in your hands to your hands themselves—is connected to and part of a deeply recursive information network. The growth of telecommunication and information technologies has transformed the tactical strategies for activism, protest maneuvers, community-building, and relating in the public sphere; and, at the same time, it has also transformed how we know the world, interact with one another, and generate what counts, at a given moment, as knowledge. The natural, social, and cultural worlds are interpenetrated by ever-denser technological systems and data landscapes. We live intensely intermediated lives. *see* **CASE STUDY 5** → 70 | 71

Ubiquitous computing, as visionary Xerox PARC researcher Mark Weiser argued, is computing that has essentially gone "invisible" precisely because it has embedded itself "in the woodwork everywhere." Ubiquitous computing—everywhere, at anytime, in everything—is possible only when high-speed networking capacities and interoperable standards allow for constant, seamless, and infinitely deep levels of information-sharing among data centers, computers, mobile devices, physical objects, and people. Nothing exists in isolation but rather in ever-denser networks of interconnection. Of course, ubiquity has a dark side: pervasive surveillance and tracking, the colonization of everyday life by information technologies, the quantification of the biopolitical sphere into ever-smaller units of analysis and monitoring, the inability or incapacity to "de-link" or "opt-out" of these technologies.

But ubiquity also allows for the massive expansion of the scholarly enterprise through a wealth of networks, information streams, and emergent communities of practice that produce and share knowledge and culture in ways that open up opportunities for participation, dissension, and freedom. Ubiquitous scholarship is marked by an ethic of collaboration and interconnection on levels that move (almost effortlessly) between the global and the local, the library and the public square, the pen and the smartphone, the millennia-long histories of humankind and the real-time feeds of the now.

The fictional case studies that follow draw from existing projects, but are themselves imaginary, offered as descriptive rather than prescriptive models for building teams, assembling the necessary resources, and launching Digital Humanities projects into the world. The case studies provide a framework for grappling with these new domains of humanistic practice.

A PORTFOLIO OF CASE STUDIES

CASE STUDY I

MAPPING DIFFERENTIAL GEOGRAPHIES IN THE NEW WORLD ENCOUNTER

In this cartographic project, techniques of thick mapping are used in combination with text analysis, data-mining, and large-corpus natural language processing. The extended project uses a participatory architecture to support annotation, debate, and repurposing of the cartographic representations and the text visualizations. Microcosmic views of the nomenclature for different geographical features are complemented by macrocosmic views of shifts in the understanding of the shape and boundaries of geographical regions.

AUGMENTED EDITIONS AND FLUID TEXTUALITY
SCALE: THE LAW OF LARGE NUMBERS
DISTANT / CLOSE, MACRO / MICRO, SURFACE / DEPTH
VISUALIZATION AND DATA DESIGN
LOCATIVE INVESTIGATION AND THICK MAPPING

Scholarly attitudes toward indigenous concepts of space and geography have changed dramatically in the last two decades through the influence of post-colonial theory. The traditional narratives of "discovery" have been interrogated, qualified, and largely abandoned. Perspectives of indigenous peoples now register within the literature, but little exists of their mapping techniques, world views, and epistemologies. Many Native American techniques for understanding geography were passed on in oral description, in myths of origin and ownership, or were indicated graphically in the most ephemeral tracings of sticks in sand or dirt. Approaches to cartography have undergone their own changes during these decades, away from what historian John Rennie Short characterizes as the story of "increasing scientific rationality" and toward maps as "social constructions, stories marked by purposeful erasures and silences."

We can now map the encounter of indigenous peoples and Europeans from different cultural perspectives, surveying incommensurate or differential geographies, explicating fundamentally distinct views of land, space, and place. This "mapping" requires careful textual analysis of the production, reception, and critical discourse around key documents in which the dialogue between indigenous and European peoples is evident. It also requires a way to produce simulations and models of a differential geography, one that would arise from the contrast of basic assumptions. For native peoples, rivers and roads were one continuous transportation route, while Europeans thought of them as land features. Europeans were focused on edges and inroads, the coastlines, harbors, bays, and means of penetrating the unknown interior. Indigenous peoples thought of land in terms of extension and activity, seasonal and tribal movements and occupation, with margins determined by social order and priority rather than physical metrics.

This project takes up the question of how, with the meager evidence before us, we can model the contrast between indigenous and European concepts of mapping at the time of early contact. Can digital means be put to the creation of an alternative view of geography and land, of spatial experience, without taking Western perspectives, epistemologies, and coordinate systems as normative? A combination of textual analysis and comparative, critical cartography will be used to explore the concept of differential geography—a mapping of space that exposes incommensurate views—and to model the changes in the historical understanding of the spaces of discovery that became the "New World."

The project focuses on several key narratives of discovery linked to European maps, some of which relied heavily on indigenous sources of information. The texts to be used include: Christopher Columbus' letters to the king and queen of Spain, accounts of Jacques Cartier's journeys into the Gulf of St. Lawrence in the 1530s, Sir Walter Raleigh's maps and accounts of Guiana from 1595, John Smith's accounts of his capture in 1607 and the map he made in 1612 of Virginia, the Codex Nuttall, Philippe Buache's *Carte Physique de*

Terreins les plus élevés de la Partie Occidentale du Canada, printed in 1754, and Aaron Arrowsmith's 1802 map of North America, which was heavily dependent on and acknowledged Native American sources.

These primary texts will be analyzed for their use of indigenous accounts and terminology. We will create a searchable corpus that will allow text analysis of key terms whose use and meaning can be tracked through the reception history of these primary documents in the critical literature. Many of these texts are already in digital form, but they have not been analyzed for this purpose. We propose to track and visualize the changing nomenclature around a cluster of crucial concepts such as space, land, mapping, discovery, contact, nativeness, and other terms to understand how the discourse of indigenous spatial understanding contrasts with that of the Europeans. We realize that the materials for authentic indigenous voices are scant, and almost all are recorded within European texts and documents. This is not a project to recover a lost authenticity, but to analyze the shape of discursive formation.

Methods

Structured mark-up, particularly the textual analysis of terms in context, will produce a study of nomenclature shifts from first contact to the present. The reception history, citation, reuse, and repositioning of Raleigh and Nuttall within the critical literature will be used as case studies since they are long, vertical studies across several centuries of use and discussion. We will also do a lateral analysis of their presence across a corpus of crucial documents, tracking usage and changing characterizations of peoples and vocabularies.

The project poses a number of technical and conceptual challenges. While some of the primary materials are in digital form, others are handwritten manuscripts whose transcription requires specialized knowledge and skill. Nonstandard spellings, shifts in language use, and errors in Optical Character Recognition will need to be checked. The question of context as a determiner of meaning for vocabulary will need to be addressed using natural language processing (NLP) in combination with structured mark-up as a method of analysis. The NLP approach will be used to identify context-dependent features of writing while the mark-up will focus on controlled vocabulary and identifiable terminology. Both can be semi-automated, and will be complemented by the use of other digital text analysis tools that can be run on the larger corpus of secondary materials to track reception history for changing terminology and nomenclature. The scale of this second phase of textual analysis would preclude analog reading methods from being used, while the first phase of textual processing answers the demands of digital technology to make explicit the judgments of the human designers of the project. Data-mining, distant reading, and close reading will all contribute to the project.

Differential cartography will be based on contrasts among European maps, verbal descriptions and terminology that can be extracted from these as having indigenous sources, and those few sources of indigenous cartography (verbal or visual) that exist in the record (e.g. Codex Nuttall is a pre-Conquest map). The task is to create cartographic simulations of an alternative worldview that does not reference European geospatial systems but has a consistent system grounded in indigenous experience, and then put these into contrast with the existing cartographic record of the "discovery" of the "new" world. Thick mapping techniques that layer historical materials in contrasting cartographic representations will allow us to present different views of the New World as a literal, as well as metaphoric, space of cross-cultural encounters. We are interested in imagining differential spatial systems and visualizations rooted in the worldviews and notions of proximity and distance, memory and community, duration and extension specific to indigenous cartographies. These are radically incommensurate with the projection and coordinate systems that are now naturalized features of standard mapping and GIS applications.

Work plan

- Identify sources for texts and maps
- Obtain permissions and digital versions
- Test the natural language processing analysis
- Create xml schema for indigenous vocabulary and nomenclature and for European geographical terms
- Mark texts and contrast search results with NLP analysis
- Search reception history corpus for usage change in terminology and vocabulary
- Create a list of cartographic fundamentals from indigenous perspectives
- Create simulations from these fundamentals
- Contrast with European maps of encounter
- Analyze the "differential" in these geographical and temporal attitudes and map them using a geo-temporal database that charts changes in attitude as shifting conceptions of space

Dissemination and participation

Poster and panel sessions at national professional meetings, postings on GeoBlog and cultural geography sites; invited response from senior figures in cultural geography and historical mapping; virtual roundtables organized as classroom events. The ongoing project will be supported by a participatory architecture that allows the mappings to be annotated extensively and also repurposed. Build on this material but extend to larger digitized gazetteer and cartography collections with emphasis on place names and cultural differences in geospatial features. Finally, develop prototypes for geospatial visualization engine that is conceptualized and structured according to the differential geographies embodied in indigenous worldviews.

Assessment

Peer review of data structure, credit for simulations, course evaluations from students on comparativist approaches in class and ease of use of analytic tools; scale of participation; ability to reveal both the limits and possibilities of interoperability with existing geospatial databases and other geo-browsers.

CASE STUDY 2

EXPANDED PUBLICATION OF A TEXTUAL CORPUS OF PAPYRUS FRAGMENTS FROM THE ALEXANDRIA LIBRARY

This textual corpus project will build a collection that links to existing repositories, makes use of certain text-based annotation platforms for debate about textual variants, performs some probabilistic natural language processing, uses a collation tool to study those variations, and creates an augmented critical edition of these fragments. Several different traditional and expanded publication models will be used to allow scholars with different profiles and agendas to present their findings in an appropriate fashion.

A new cache of papyrus fragments has been discovered in Alexandria, Egypt. Though considerably damaged by neglect and wear, these fragments promise to answer some long-standing questions about the spread of the Phoenician script and dates of its adoption across North Africa, particularly the coastal regions to the west, and its possible dissemination along trade routes into India. Some surprising features of these papyri make clear that they were recycled several times in the course of their use. Many are palimpsests, and some have several layers of script in varying degrees of legibility.

A host of different imaging technologies and digital platforms for integrating the data collected from the papyri are currently available. One major part of this project will be to repurpose some of the techniques that have been used successfully in other projects in Western Semitic epigraphy. However, the language in these papyri is not limited to Semitic tongues, and to the surprise of the librarians involved in the discovery, several as-yet unidentified languages seem to have been making use of scripts whose use and spread had been thought to be well-documented. One scholar has suggested the presence of Indo-European roots in the organization of the linguistic structure, which would suggest earlier contact with the Indian subcontinent and more dramatic cultural diffusion and influence than has previously been thought. Before any natural language processing can be done on the texts, they have to be deciphered (because of their poor condition). NLP techniques for ancient languages are in experimental stages, especially with fragmentary sample sets. Some speculative and probabilistic readings of the papyri and of the texts will be used.

The research problem is to identify the language groups represented in these papyri, match the script forms and use with the known corpus of Semitic epigraphy, and track the variants in a database that can support data-mining and text analysis across versions, translations, and script forms. A side benefit will be the creation of a digitized corpus of the papyri. One of the difficulties is that a major figure in early Indo-European languages is elderly, ill, and unable to travel so that his input will have to be done entirely on digital surrogates. A platform for annotating and tracking his contributions will have to be built or repurposed from existing platforms. This is an incentive for involving a team of mid-career and younger scholars whose formulation of problems of linguistic change and diffusion will also be essential. They are demanding an augmented publication format that will allow their work to be published rapidly, with a short peer-review cycle and with various scales of intellectual contribution, links to other existing corpora and repositories, and even an agonistic spirit of gamesmanship.

Methods

To do this project effectively, some crowd-sourcing of the translations, decipherment, and editing of the documents seems like a viable possibility. Statistical methods for doing large corpus analysis and comparison will also be essential. Thus, both close-reading and distant-reading techniques need to be involved in the study of the texts, the artifacts, and the scripts. The first phase of the project will require extensive integration of the imaging and digitization, with all uncertain signs or graphic elements marked so that the guesswork part of the project is conspicuously noted. Using the cultural analytics platform for display of large numbers of artifacts as well as pattern recognition software, similarities in script forms will be used to pinpoint linguistic similarities. The text translations will remain fluid, with variants and disputed elements conspicuously marked. Publishing these bits and pieces on a regular basis will be essential if any crowd-sourced work is going to occur. The senior scholar has asked that his interpretations be given a separate layer for presentation online so that his work can stand alone and be scraped off for later publication in print format.

Work plan

- Identify imaging techniques and sources of equipment
- Establish partnerships for shared access
- Create image files and test integration and comparison techniques
- Test probabilistic methods of text analysis for fragmentary data sets
- Identify translations of source texts where appropriate
- Link to existing repositories and online translations
- Test annotation and version-control platforms
- Test the cultural analytics and pattern-recognition software
- Put peer-review system into place for short-cycle contributions
- Create a platform of publishing and crowd-sourcing translation, editing, and decipherment
- Continue iterative process of imaging, translation, and decipherment
- Do sequential publishing of the findings in the form of an augmented edition that contains links, comparisons with existing corpora, and other versions of the texts

Dissemination and participation

Create a Twitter feed and RSS feed to publicize the project and engage participants; publish a beta version of the project in a print-plus mode online and establish a workflow to repurpose this content for traditional publication; crowd-source the translation and comparisons as well as the decipherment; augment the edition on an ongoing basis as scholars in the field indicate points of connection or comparison with other existing papyri, texts, or fragments of ancient scripts and languages. The project will be linked to major repositories in the Near and Middle East, Europe, and the United States by using an aggregation engine to allow for a larger statistical sample for investigation and comparison. The bridging of traditional and new modes of scholarly engagement through distributed knowledge production approaches will allow the senior scholar to work effectively with younger scholars and allow for crowd-sourced input without collocation.

Assessment

The technical, intellectual, and cultural/institutional aspects of scholarship are interdependent. Success will be gauged in part by the extent to which the decipherment is completed and legibility for various layers established with credibility through the imaging and textual analysis. Another measure of success will be the number of contributions that enable links to existing digitized fragments and/or translations of ancient scripts.

CASE STUDY 3

AUGMENTED OBJECTS & SPACES: JEWISH RITUAL OBJECTS IN DIASPORA

This project in critical curation and the augmentation of objects with commentary will resituate religious objects though a multi-modal approach that captures ritual practices across the time and space of diaspora. Spearheaded by a museum, the goal of this project is to produce an animated archive of cultural materials attuned to questions of provenance, use, and scholarly interpretation.

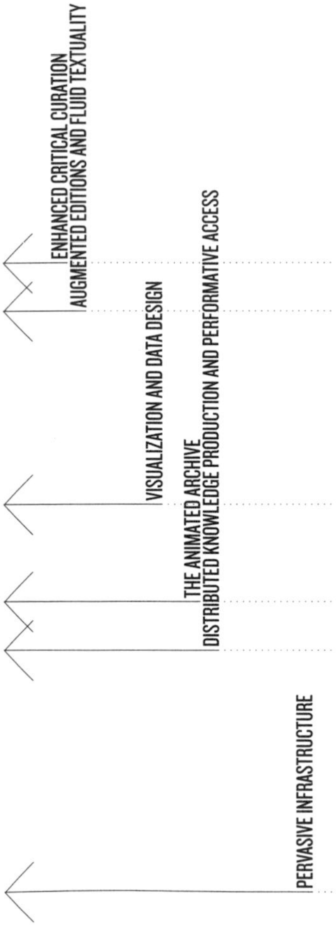

A university museum has an extensive collection of costumes, ritual objects, and recordings of ceremonies from Jewish families in Poland. These were obtained across several generations and do not all have fully documented provenance. Some were the work of early 20th century anthropologists who brought objects, photographs, and artifacts back from the field. Some are materials that were part of a large collection developed by a mid-century alumnus who donated them to the museum on the condition that the materials be used for teaching and public education about Jewish art and culture. Other materials were acquired through a fund established for the promotion of the study of the Jewish diaspora and were bought at auction or through reputable dealers by the museum curator. But some materials are of uncertain origin, and have been the subject of controversy, since they may have been looted, stolen, or smuggled out of the countries of origin, possibly as the result of Nazi appropriation. The museum has started a major initiative to make use of its Jewish ritual object collection and create a series of public programs, research opportunities, and curricular initiatives, as well as produce some permanent exhibits. In order to do this effectively, the director has determined that a digital approach based on augmenting the objects with multifaceted information displays will be the most effective way of addressing the ethical and intellectual dimensions of this cultural legacy.

While much of this material has been cataloged, not all of the descriptive information about the works has been put into digital form. Some of the earliest material was entered into the registrar's ledgers in handwritten form, while the most recent metadata conforms to the Getty's standards for Cataloging Cultural Objects. While the museum staff would like to standardize metadata for the purposes of managing the collection, they do not want to lose the important record of earlier approaches to the classification of artifacts. The idea of displaying different subcollections within the larger whole also suggests some interesting historical narratives about the development of diasporic anthropology and cultural studies of ritual objects. Finding an effective way to display different interpretive approaches is crucial.

The artifacts in the collections range in size and scale from tiny mezuzot to tefilin, prayer shawls, Torah coverings and a fully rebuilt antique ark and bimah. Photographing these works for digital presentation also poses some challenges. Thinking through the organization of images to show multiple views, to facilitate detailed as well as overall study of the objects, and to allow for research as well as interpretive exhibits will take some serious repository building and design. The curators want to avoid any kitsch or special-effects approaches and also do not wish to create fictional spaces for actual artifacts.

ENHANCED CRITICAL CURATION
AUGMENTED EDITIONS AND FLUID TEXTUALITY
VISUALIZATION AND DATA DESIGN
THE ANIMATED ARCHIVE
DISTRIBUTED KNOWLEDGE PRODUCTION AND PERFORMATIVE ACCESS
PERVASIVE INFRASTRUCTURE

One part of this collection came from a synagogue in Poland that was destroyed in World War II. But the site has been excavated, and there are extensive field notes and site photography to accompany the artifacts. These objects could be resituated through the narrative of the dig and accompanied by a story of the excavation. Other objects are of dubious provenance, and so need to be presented in a manner that allows serious scholarly engagement with their history and forms. A significant number of artifacts are known to be stolen, and finding descendants of survivors who can claim them is important and will be a part of the outreach supported by distributed access to the collections. Descendants of survivors will need to show appropriate credentials and be vetted before they can search and annotate the archives and repositories. Display of these materials may need to be limited, but research on them needs to be supported digitally so that some of the scholars best positioned to do this detective work can access them. The museum will collaborate with restitution groups which provide legal expertise and advocacy on behalf of survivors and their families.

Perhaps the most challenging materials among the collections are recordings of ritual practices that were never meant for public display. Some of these recordings were obtained surreptitiously. Others were obtained under very carefully worked out privacy agreements and intellectual property negotiations. Creating an environment that respects these agreements or goes further in using the museum environment to educate the public about the restraints on viewing seems essential in today's critical frameworks.

The digitization of these collections and creation of critical exhibits for public programming, education, and research is the focus of this project. The goal is to create exhibits that augment the objects and artifacts by exploring these many intellectual, critical, ethical, and political dimensions. Relating the artifacts to the geo-temporal history of diaspora is one component of the exhibit. But community testimony and archival materials that provide demographic data are also crucial contributions from which to generate display. The ultimate goal is to present the history of diaspora told through the movement of things and the rituals around their use. Creating a network analysis and information visualization will be one part of this presentation. Another will be the attempt to situate all of the artifacts within practices. Thus an artifact will never appear as a single thing on display. No artifact will be an autonomous object with a single text label. Instead, all objects will be accompanied by a digital matrix that exposes provenance questions, communities of use, and historical information about each as well as information on its acquisition, transmission, and debates about the ways it should be displayed and interpreted. In other words, the display of augmented objects will refract them along multiple lines of inquiry and interpretation that invite scholarly and critical engagement with an animated archive of materials.

Work plan

- Inventory the objects and artifacts
- Do an assessment of the metadata and cataloging protocols
- Create a set of crosswalks and schemes for description of the objects
- Address the multiple representations of the objects in existence and those to be created through the process of photographing or scanning
- Identify a content management system appropriate for museum management
- Test various network analysis tools and visualizations to display the movement of objects through time and space
- Modify the system so that it is customized for appropriate workflow and use
- Consider the administrative issues of permissions, access, and use of digital materials
- Consult appropriate scholars and authorities on legal and ethical issues around these materials
- Create an academic and community advisory board for ongoing review of ethical practices
- Create proof of concept demonstrations of augmented object displays that contain multiple viewpoints and artifactual histories
- Develop an appropriate permissions system and demonstrate the ability to create tiered levels of access and use for various audiences
- Design a method for processing input from professional and amateur scholars
- Consider the ways to engage stakeholders in the larger questions of cultural ownership

Dissemination and assessment

Create text labels, commentary, and debate that carry an author attribute; design a system of searching and indexing according to author; record and display relevant debate trails generated by objects; create a public forum in which these debates are edited or represented for study; engineer an app for mobile devices that allows input from contemporary sources.

CASE STUDY 4

VIRTUAL RECONSTRUCTION OF AN AFGHAN REFUGEE CAMP AS A SITE FOR CULTURAL MEMORY

This project repurposes the technology of online multiplayer games to create a virtual community of testimony, witness, recovery, and social bonding. A spirit of joy and community-building is present as the shared repository of memories—photographs, some video materials and audio tapes, as well as letters, diaries, journals, and other materials—is being used by a younger generation to create a shared history through a series of mash-ups in which nuclear family histories become the common property of extended "families" through database documentaries and remix storytelling. This shared history may promote political activism, but also may become a target of unintended surveillance. Use of avatars and assumed identities is standard practice, and sensitive materials are likely to be part of the repository.

VISUALIZATION AND DATA DESIGN
THE ANIMATED ARCHIVE
DISTRIBUTED KNOWLEDGE PRODUCTION AND PERFORMATIVE ACCESS
HUMANITIES GAMING
REPURPOSABLE CONTENT AND REMIX CULTURE
PERVASIVE INFRASTRUCTURE

A professor specializing in politics, one in architectural history, and another in performance studies have been gathering material for a collaborative research project that would allow them to create a virtual model of one of the largest refugee camps that came into being after the Soviet invasion of Afghanistan in 1979. Interest in the site and its inhabitants has been spawned by recent events, including the pullout of American troops. Many of the children born in that camp grew up outside of Afghanistan—in Pakistan, Iran, India, and elsewhere throughout the region and beyond. An international organization interested in "virtual" repatriation is looking at patterns of diaspora, assimilation, and cultural memory. The idea of using a virtual reconstruction of the camp as a point of shared experience touches many nerves. The site itself, though still in existence, is in sensitive territory, difficult to access. But photographs taken by a U.S. solider have been smuggled to the organization, Jalozai International, and offered to the U.S. academic team. A mobile phone application that repurposes these photographs to create an augmented reality experience of the original site has gone viral.

The site needs to be re-created virtually, but should be as accurate as possible with respect to the layout of the original camp. The group is working with a refugee community organization, which is in touch with a worldwide diasporic network of displaced persons and refugees. This community organization is eager to participate, as much as possible, in the creation of a virtual environment that could serve as a theater of reconciliation, testimony, memory, and commemoration. The United Nations High Commissioner for Refugees helped with repatriation, while the Pakistanis issued ID cards to all Afghani people living within their borders. For political reasons, many persons slipped through these official programs.

Much research has been done, and hundreds of hours of interviews have been logged. Descriptions of the camp from firsthand accounts, and from photographic and drawn images, have been gathered. The reproduction of everyday life in the camp would be made in virtual form in a simulation lab. The integration of the stories and eyewitness accounts and the creation of a fully immersive environment represent the next phases of research for this project. The questions surrounding the use of the environment and the quasi-game-like virtual world it suggests are all beginning to raise some concerns in the university and in the community. Creating a way to allow active participation and contributions from the community without trivializing the trauma of those who experienced the camp firsthand is one problem. Keeping fake testimony and malicious content from appearing is another. Protecting sources is yet another. However, all involved are interested in using theater and performance art as a way of engaging with recent history. They want to treat the camp not only as a historical site, but as a living memory that has to be engaged directly through imaginative experience if it is to be fully understood. Members of the younger generation are engaging in the creation of remix narratives and role-playing games based on materials in the repository.

The goals of this project are to create the immersive simulation that allows for performance of recent trauma in an environment that may or may not be able to be controlled. Some simulations and predictive models are being built into the system, particularly those that use complex adaptive systems modeling techniques, and these will be used in dialogue with live user contributions to monitor emerging trends in the environment. However, any hint of surveillance or control will have disastrous results, and the simulations need to be fully transparent to all participants. The performance studies professor has been working with the interface designers to produce some avatar representations and thumbnail theaters that show possible scenes and probable story lines among live participants. Creating an interface that allows multiple users to participate actively in a multi-person performance while also making use of historical materials and documents will require careful scripting and guidelines. A polymorphous browser that displays materials differently depending on how parameters identifying the user are set is in beta.

Methods

Identify the available software for creation of a virtual site with social media and participatory capabilities, or consider making this site in Second Life or another virtual world application. Get information on issues of security and privacy if a third-party platform is used. Make sure the site can be used with mobile apps as many of the participants will not have cable connections, but will access the materials through their cellphones.

Work plan

- Establish communication with appropriate international organizations
- Develop collaboration with the refugee groups and their leadership
- Create a beta version of the virtual site
- Invite a small group for user testing
- Perform iterative user testing based on initial results
- Build out the virtual environment
- Create a public forum for input
- Create avatars to protect privacy and identity as appropriate
- Document use and participation
- Create cross-referencing tools for tracking shared information and memory

Dissemination and participation

Engagement of the UNHCR and dialogue with the Jalozai International leadership is crucial, as is ongoing support of the university where the project is housed; plans for a small working group to meet in Iraq and another in Pakistan will facilitate direct contact with academic team members; YouTube presentations and virtual encounters are also planned, as is a series of performances in the virtual environment. These will be publicized as real-time events in the virtual space. Scholarly publication of findings will take various forms, including but not limited to traditional conference presentations and publications supplemented by digital collections and archives of the project and its materials deposited in the university library.

Assessment

Assessment will be ongoing; monitoring the participation and reaction of participants will be crucial to safeguarding privacy and gauging comfort levels as well as the effectiveness of debate, dialogue, and documentation; getting solid documentation of the contributions and testimonies is essential.

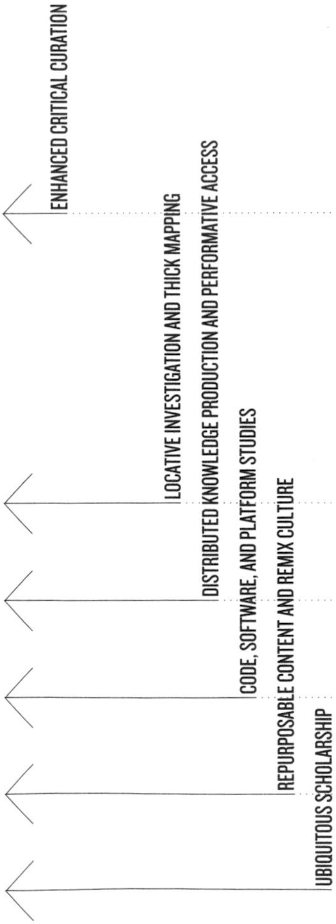

CASE STUDY 5

MULTI-AUTHORED LOCATIVE INVESTIGATION OF THE ZENON HEADQUARTERS AND CORPORATE ARCHIVE

This project aims to design a prototype tablet application that interweaves three components: a collaboratively developed body of interpretive research, an archive, and an architectural heritage site. The team will build a content management system for the online generation of a multi-authored essay and a digital archive that will be keyed to the physical structure of the heritage site. User-visitors will be provided with a navigation system that allows for multiple entry points and pathways for on-site and off-site access as well as for "reading" this "published" work. The prototype will be employed to beta test a new way of accessing information, interacting with knowledge, and experiencing data and research in physical and virtual space.

ENHANCED CRITICAL CURATION

LOCATIVE INVESTIGATION AND THICK MAPPING

DISTRIBUTED KNOWLEDGE PRODUCTION AND PERFORMATIVE ACCESS

CODE, SOFTWARE, AND PLATFORM STUDIES

REPURPOSABLE CONTENT AND REMIX CULTURE

UBIQUITOUS SCHOLARSHIP

An archive of documents from the Zenon Corporation, whose headquarters occupied a historical building designed by Louis Sullivan, was discovered during recent restoration work. Stored in a safe in the basement of the building, the documents have now been moved to the University of Illinois, Chicago, where they have become the focus of interdisciplinary study. The archive contains copies of business correspondence, internal memos, meeting notes, drafts, minutes, calendars, and personal notes. Some are handwritten, some are in shorthand, and some are hand-edited drafts. Others are typewritten documents, many on letterhead featuring the building with its original facade elements, including downspouts and decorative features that were removed in the 1970s. The documents are from the 15-year period (1945-1960) during which Zenon, contracting with UIC, was developing a secret project for the U.S. military to build a distributed computing system for defense that many cultural historians believe was an early version of the networked computer.

The cast of characters is colorful. Zenon's president was a Harvard-educated scion of an old Chicago family, and the Sullivan building had originally been built for the home offices of his grandfather's insurance business. Plans for the project were leaked by his secretary and mistress, whose connections to gangland South Side mobsters seemed to hark back to Depression-era speak-easies. Scandal and intrigue dogged the Zenon project, and the secretary's body was found at the bottom of an elevator shaft, a crime concealed for a half decade. The documents also open up a window into the culture of the corporation and the research university during those years as well as the relationships among employees as mediated by the writing technologies of the time (pen and paper, stenographic machines, typewriters, shorthand).

An interdisciplinary team of scholars is working on the reconstruction of the Zenon project within a "history of the future" framework. It seeks to publish a "multidimensional essay" built around a core set of archival documents—the correspondence between an executive and his secretary—in order to explore a novel publishing model that allows for the building of connections across media, as well as across digital/physical boundary lines. One team member is a cultural historian of organizations; another is an expert on the history of network architectures. A third is interested in the "gendering" of spaces and forms in corporate architecture and, working with the team, seeks to develop an augmented reality app to reconstruct the original Zenon building at its site. Her work is funded by the Chicago Architectural Association and the Chamber of Commerce, both of which wish to integrate scholarly content into tours of the city's sites available on mobile devices. She proposes structuring the interpretive work being carried out, as well as the Zenon archive itself, as a function of the building's layout and spatial organization such that the site visitors are able to "see" events unfold at set points in the entry, lobby, elevators, and hallways, thereby fulfilling the needs of both funders. But the project must also be designed so that off-site visitors can experience the full interplay of archival materials, interpretive research, and architectural features.

Working alongside the team of scholars is a technical team comprised of an interface designer who specializes in digital corpora, a computer science doctoral student working on information structures and knowledge models, a designer who is an expert in CAD systems, modeling, and architectural rendering, and an adjunct professor who spends half his time as a technology developer creating tablet apps. They will play the lead role in creating an application that allows: a) on-site users to move and rotate their tablet devices to navigate 3-D models of Sullivan's building framed by and animated by texts, annotations, commentary, and archival documents; and b) off-site users to replicate this experience on their own tablet devices in clear and meaningful ways. This prototype will allow scholars from across the humanities to test the viability of multidimensional formats as publishing platforms.

The software challenge is three-fold: First, to create a content management system that allows the three scholars to collaboratively generate texts, metadata, annotations, and images; interweave them with the Zenon archive; and embed them in the architectural models. Second, to create an intuitive, user-controlled navigational interface that supports both on- and off-site movement through the combined content governed by the spatial structure of the building, semantic features of the content, the user's prior navigational choices, and user settings regarding the density of data layers and frames. Third, to build richly textured digital representations of the interiors and exterior of the Sullivan building that are appropriately keyed to the various data layers that make up the research project.

Work plan

- The entire team works together to develop the information architecture and content schema
- The developer customizes or creates a bespoke content management system to be used in the writing of the three intertwined "essays"
- The scholars generate the writing, annotations, diagrams, images, and other content, along with metadata and links
- The technical team devises a database architecture for the selected portion of the Zenon archive that will work seamlessly with the content management system and the augmented reality app and its architectural models
- The designer and developer iterate and test interface designs, user interaction, and navigation for both on-site and off-site access
- The team goes back and forth between scholars generating material and the designer and developer demonstrating the application until an alpha version is ready for user testing
- Alpha testing with users both for the performance of the application but also for the reading and interpretation of the content
- Develop a website for feedback and user participation
- Iterate a beta version based on feedback and refine for public release

Dissemination and participation

The application will be made available for download from online stores at no charge. Its availability will be broadcast through social media and will be promoted by the supporting institutions in a version that is "tuned" for nonspecialist audiences. The project website will continue to be a site of conversation in the hope that other teams of scholars might contract with the technical team to undertake new and extended versions of the project. The results of the project, including user feedback, will be documented, analyzed, and shared through conventional venues, such as conferences and journals, from each of the team member's fields.

Assessment

Assessment will be built upon three points of engagement. On-site and off-site user testing from the standpoint of human-computer interaction will be a part of the development process. User metrics will be generated through social media rankings, online store rankings, number of downloads, and direct feedback. Critics and scholars will be asked to review the application from a variety of vantage points: as a model for scholarly publishing and for the merits of the scholarship of the three contributors.

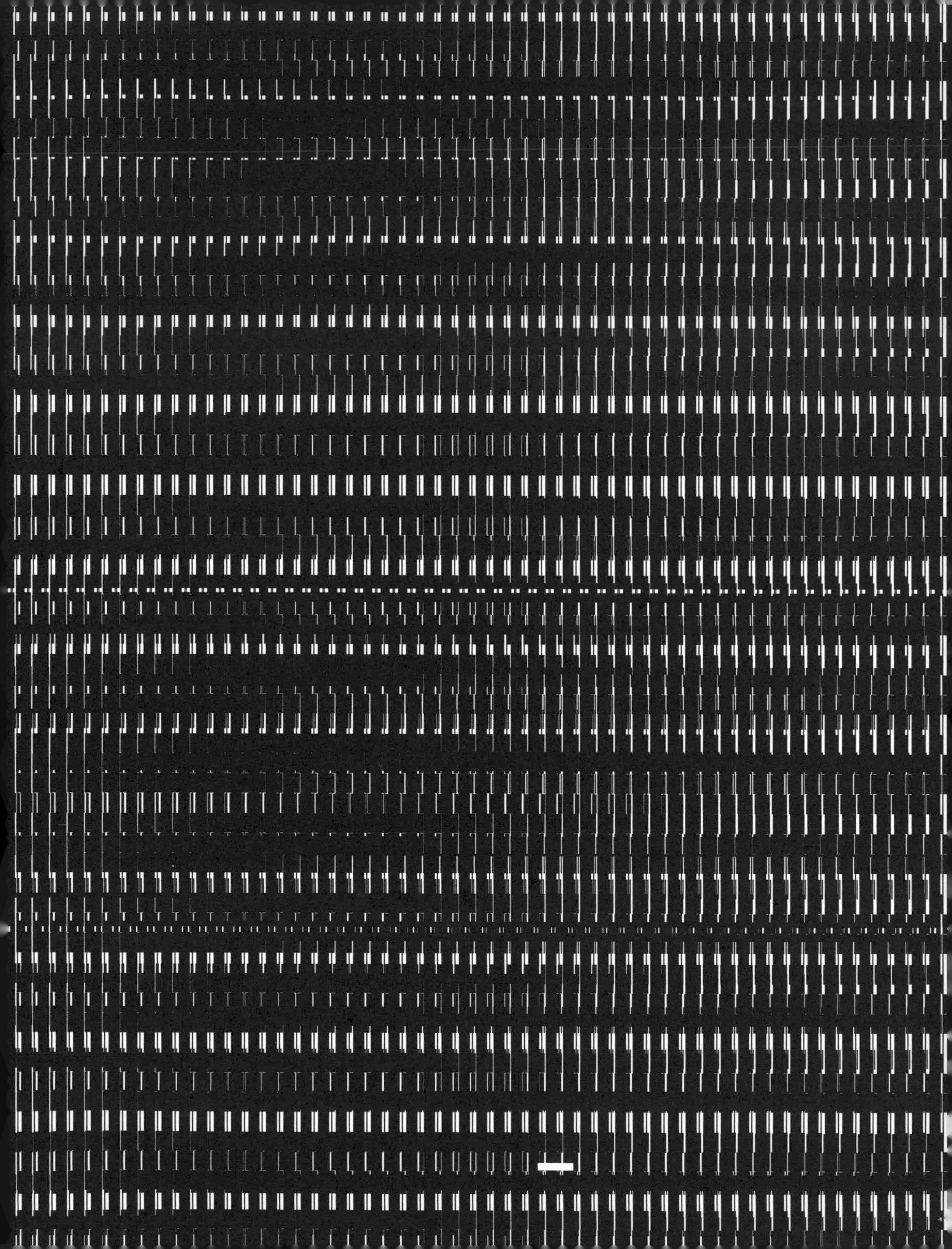

3. THE SOCIAL LIFE OF THE DIGITAL HUMANITIES

DIGITAL HUMANITIES ENGAGES A WORLD OF LINKED AND LIVED EXPERIENCES.

BECAUSE NETWORKS CONNECT US, THEY ARE SOCIAL TECHNOLOGIES. AS SCHOLARSHIP MOVES FROM THE LIBRARY AND THE LECTURE HALL TO DIGITAL COMMUNICATION NETWORKS, IT TAKES ON EXPANDED SOCIAL ROLES AND RAISES NEW QUESTIONS. NEW MODES OF KNOWLEDGE FORMATION IN THE DIGITAL HUMANITIES ARE DYNAMICALLY LINKED TO COMMUNITIES VASTLY LARGER AND MORE DIVERSE THAN THOSE TO WHICH THE ACADEMY HAS BEEN ACCUSTOMED. THESE COMMUNITIES INCREASINGLY DEMAND AND DELIGHT IN SOCIABLE INTELLECTUAL INTERACTIONS, IN WHICH CRITIQUE MANIFESTS AS VERSIONING, AND THINKING, MAKING, AND DOING FORM ITERATIVE FEEDBACK LOOPS.

A DIVISION emerged over the course of the 20TH century that separated humanities knowledge into study and analysis on the one hand, and practice and application on the other. The former is characterized by criticism, hermeneutics, and close reading, almost exclusively undertaken by a single author who works to articulate a highly defined problem in a specific discipline. The latter is rooted in design, collaboration, and performance, often stretching across media and involving multiple agents, producers, and authors. Thus, the creative energies of the arts come to be seen as distinct from the "serious" practices of criticism, analysis, theory, and history. In other words, the process of "how" became separated from the content of "what."

This division helps us understand the ways in which the diversity of humanities knowledge is regularly (and not always unfairly) stereotyped as a dry, rarefied, canonized set of objects, disciplinary practices, and media forms. The "how" requires attention to design, format, medium, materiality, platform, dissemination, authorship, and audience, things that are all taken for granted or assumed to be implicit, value-neutral, secondary, or even irrelevant when scholars turn over their manuscripts to a university press. But there is nothing neutral, objective, or necessary about the format of a book, the space taken by a page, the medium of paper, or the institution of a press. In fact, the "what" is shaped by the "how" in a profoundly recursive, process-oriented manner. When print artifacts are no longer the primary medium for knowledge production in the humanities, norms begin to change and the "how" of design reasserts itself at the core of every "what." In embracing such a transformation, the Digital Humanities not only takes on a new set of disciplinary and technological tasks, but also a world of linked and lived experiences that are at once social and epistemological in character.

This chapter focuses on the social aspects and societal impact of the Digital Humanities. It begins by analyzing the economies in which humanistic knowledge was created in the past before turning to how open-source models, information technologies, and social media have reshaped contemporary practice, promoting social transformations that affect the reach and relevance of humanities work. One such impact is the invigoration of collaborative authorship and the growing significance, in turn, of what will be referred to as the "curation of knowledge." Publishing, always a social act, becomes ever more so in the Digital Humanities, challenging academic presses and university libraries to stay true to their mission of promoting excellence while reaching out to expanded publics. The chapter concludes with a description of the *hedgefox*: the new type of student that a Digital Humanities education could and should produce.

Open-Source Knowledge Economies

In order to understand the transformation of scholarship and scholarly practice in the Digital Humanities, we can contrast two different economies: The first is that of the Industrial Revolution which shaped the establishment of modern humanities disciplines and departments within Western universities; the second is the globalized economy of the networked information age, the economy of the Digital Humanities. It is here that we can discern a critical difference between an economy of knowledge production tending towards scarcity, centralized control, hierarchy, division of labor, property, and proprietary systems versus an economy of knowledge production tending towards abundance, decentralization, peer creation, creative commons, and open-source models.

The fact that, by its nature, Digital Humanities encompasses both an academic and a social life expands the discussion well beyond the technological. Central to the transformations of the 21ST century is the movement from closed- to open-source cultural production. Open-source culture possesses a multitude of facets and definitions, comprising many of the attributes already discussed: collaborative authoring, multiple versioning, flexible attitudes toward intellectual property, peer contributions, access to multiple and multiplying communities, and overall patterns of distributed knowledge production, review, and use. Open-source cultural concepts had their origins in the software development community decades ago, when dedicated independent programmers revolted against the decision by computer manufacturers to sell software where the source code was "closed," meaning it was impossible to change, improve, or adapt to the user's needs. The free and open-source movements started to create alternate operating systems and programs that users could contribute to, improve, and send back out to the developer and user communities. Richard Stallman, one of the leading lights of the free software movement, once referred to open-source projects as "technical means to a social end."

The growth of networks facilitated and accelerated open-source production, allowing the globally connected to ship code from developer to developer, upload new versions, and check for flaws and improve performance. Proof soon came in the form of the wildly popular open-source Linux system, originally developed by a Finnish university student named Linus Torvalds and then expanded by a vast community of like-minded programmers around the word. The maxim of open-source software developers—"given enough eyeballs, all bugs are shallow"—was a fresh way of thinking about how robust, stable systems could be the product of multiple, autonomous hands rather than of centralized, top-down, proprietary models of development.

As Stallman anticipated, what began as technical became social, and ideas of free and open cultural production began to percolate through society. What this meant was that users wanted to be able to mine networks and systems for parts and even wholes that they would then be free to refashion, remix, and re-create according to their wants and needs. This was, and remains, a utopian prospect, in large part because open-source runs up against inherited notions, values, and rules regarding intellectual property. **For if code—or any cultural product—is produced by a distributed network of sometimes unknown creators, how is it to be regulated? Who is held accountable? Who owns it?** Open-source has come under attack from centers of power because it challenges the very intellectual property rights that sustain many dominant players within the global economy. Open-source also encounters opposition from communities and institutions that are committed to limited, calibrated, or stratified forms of access to cultural knowledge. Indeed, it seems more than legitimate to critique open-source's utopian universalism inasmuch as all knowledge or cultural materials cannot be shared on equal terms. Tracking the provenance of materials that reside in a given cultural institution can expose histories of violence, plunder, or genocide just as well as it can expose heroic acts of rescue and preservation. So it is incumbent upon contemporary humanists to embrace a thoughtful, critical attitude toward open-source resources. No single or rigid approach to cultural property suffices. Digital diversity means recognizing fundamental differences as regards technological platforms and the uses to which materials are put.

To fully understand the specificities of our current moment and the prospects for the future, we need remember the social life of information in the industrial era, when worth tended to be defined by scarcity. Trade secrets were feverishly guarded; access to the means of production—not to mention media and information technologies—was controlled; participation was limited by decidedly hierarchical functions and divisions of labor; and property was owned, protected, and sold in an economy that reduces value to supply and demand. Though there were notable exceptions to this model of scarcity—the establishment of public education from kindergarten through graduate schools, the great philanthropic initiatives, the building of public library systems—much of our legal and economic system is still predicated on scarcity and narrow definitions of ownership as a driving force. By contrast, the networked information economy, at its best, promises openness, sharing, and common platforms for information exchange. Access to the means of production in the information economy is dramatically opened up, rendering the bar for participation low enough for nearly everyone connected to the Internet. Peer-to-peer sharing and open-source models of production transform "property" into something created, edited, and monitored by the ever-expanding public but ultimately owned by no one. Many defining aspects of the networked information economy are based on abundance and the copiousness of the digital copy, which in turn is based on the cooperation and openness that characterized the early years of network development.

No clean break exists between these two economies and elements of complexity and contradiction appear within each. But forces are sure to continue to vie for control in an era of (seemingly) seamless networks, open platforms, and global exchange. For Duke University Law School's James Boyle, the real danger to the "commons of the mind" is not unauthorized file-sharing but "failed sharing." Rather than participating in the corporate panic about intellectual property theft, Boyle argues that we should be concerned instead about the enclosures and strictures placed upon the world of the creative commons. This is a debate that rages on, and digital humanists will increasingly be called upon to provide intellectual capital in the struggle with the monetary capital of some corporate titans as they wage a legal and cultural battle to regulate, protect, and monetize the intellectual property set free by the World Wide Web, the global merging of networks, interactive technologies, and digital cultural production.

Social Transformations and Technologies

As we have seen, the industrial economy which typified production and defined social relations over the past two centuries has been transformed over the last two or three decades into what Yochai Benkler terms the "wealth of networks" that characterize the decentralized and open information economy. With the growth of the participatory Web and social media technologies—what many have called "Web 2.0"—we have seen the emergence of an economy defined by social structures, modes of production, and cultural formations that alter the way information is produced and exchanged, enabling a global and networked world of decentralized sharing, collaboration, and diffusion, with the caveat that it also creates the conditions for violent backlash and newer forms of surveillance and control.

What makes social technologies different from unidirectional technologies like broadcasting? First, the bar to entry for using contemporary social technologies is remarkably low. Provided access to the Internet (which, to be sure, is not a given), anyone can tweet messages, upload pictures and/or video, post blogs, and download a wide range of media content. Secondly, social technologies are indeed "social," which means that they are communal, community-generated and community-generating. To socialize is to follow, to participate in, and to associate with—a structurally different way of conceiving of technologies than, say, instrumental uses to "do" certain tasks, which was the original model of the personal computing revolution, or private uses of technologies to "limit" access, as in the commercial obsession with intellectual property rights. Social technologies create social communities and public cultures that complicate and often transcend boundaries based on geography, age, class, ethnicity, gender, and so forth. And thirdly, social technologies have histories that parallel, influence, and give shape to human social structures and societies: Writing *qua* writing is a social technology; the postal system is a social technology; telephones, email, and IM are social technologies, precisely because they create interconnections and networks of communication, dialogue, and interaction that enable and give rise to relations that form the basis of societies.

Nowadays, it is common to celebrate the democratizing and decentralizing possibilities of social technologies, but we need to consider social media—like all technologies—from the standpoint of the dialectic: They enable broad-based participation; the bar to participate is remarkably low; and they produce global diffusions of information, often through precarious portals that would otherwise squelch voices. Would Egypt's triumphant Tahrir Square have turned out like China's Tiananmen Square had the last two decades not seen the proliferation of cellphones, Twitter, and Facebook? At the same time, social technologies are also beholden to an array of political and corporate interests which have amassed large and complex data sets relating to every aspect of our behavior in order to create perfect consumer profiles, track identities, and enable ever-greater forms of surveillance and population control.

All over the world, authoritarian regimes have turned social media to their advantage much as they manipulated prior media. By intercepting postings and passwords, by asking users to agree to new security certificates and engaging in other coercive techniques, and by passing off propaganda as "spontaneous" participatory content, governments or partisan groups can amass troves of data to identify dissidents, abuse power, or create smoke screens regarding public opinion. We raise these issues with respect to the Digital Humanities to underscore the fact that social media not only enable democratic ends but can also make possible domination and subjugation. So, as much as we celebrate the global proliferation of networking, it is important to bear in mind that network technologies do not inherently promote democratic values and community-building. They also create the conditions of possibility for violent backlash, community surveillance, and possibly even genocide. After all, the railway system—the paradigmatic networking and social technology of the 19TH century—not only enabled transnational movement and the birth of the global industrial economy, but also provided the technical means for efficiently deporting whole populations to face their murder in the 20TH century.

The socialization of interior life and the restructuring of individual subjectivity in the face of constant communication exchange may yet produce long-lasting changes in the concepts of public and private space, security and privacy, identity and community. Swarm behavior and collective absorption into real-time activities have already created new models of rapidly organized and mediated communities. The interpellation of interior life into the networked environment is unprecedented, and the fate of the individual voice hangs in the balance. At the sci-fi end of speculation, collective thought forms seem to lurk on the edges of our horizon.

Reach and Relevance

What does all this mean for the Digital Humanities? First of all, the humanities are one of the key places to which we naturally turn to understand, analyze, and evaluate the social and cultural significance of any technology, to interpret its value, its dangers, and its possibilities. This, we contend, makes the work of the humanities more critical than ever as new social structures, economic models, cultural forms, value systems, and forms of selfhood emerge, rendering the "human being" decidedly more motile, diffuse, and even fragile. Broadly speaking, since the Digital Humanities studies and explicates what it means to be human in the networked information age, it expands the reach and relevance of the humanities far beyond small groups of specialists locked in hermetically sealed conversation. The scope and scale of the Digital Humanities encompass a vast archipelago of specialized domains of expertise and conversation, but also open up the prospect of a conversation extending far beyond the walls of the ivory tower that connects universities to cultural institutions, libraries, museums, and community organizations.

In fact, the notion of the university as ivory tower no longer makes sense, if it ever did. Instead, our ideal is that of the university as nodal point within a fluid, porous, and dynamic landscape. (Even medieval universities gave rise to their own networks of social life and publishing, albeit on a different scale and with manuscript copies and lectures as their defining media.) The social life of the Digital Humanities builds upon that landscape by making possible a networked information economy characterized by collaborative authorship and design, the social production and dissemination of knowledge, writerly authorship models predicated on multiplicity and versioning, participation in the expanded public sphere, and institutional and non-institutional venues for designing, sharing, commenting on, critiquing, and—perhaps most important of all—engaging with this knowledge.

Altered Modes of Authorship

"What is an author?" is a question that has long been central to humanities scholarship. Traditionally conceived, authors are individuals who compose on their own but write in dialogue with a community of peers and a received tradition. They—poets, philosophers, historians, novelists, dramatists, and so forth—create worlds through the written word. As Aristotle's distinction goes: We read historians to know the world, to understand what happened; we read poets to imagine what might be, to inspire new worlds into being. Digital humanists share traits with both historians and poets: We are engaged with "worlds past" and also with worlds that are not yet. But digital humanists imagine the past and the future in ways that fundamentally transform the authoring practices of poets and historians, using new sets of tools, technologies, and design strategies. For digital humanists, authorship is rooted in the processes of design and the creation of the experiential, the social, and the communal. We no longer imagine authorship as autonomous work or as the labor of a solitary genius (something that, to be sure, critical theory has been chipping away at for decades). Instead we think of the harnessing and expressiveness of the creative energies of an ever-expanding, virtually boundless community of practitioners.

The question is no longer "what is an author?" but what is the author function when reshaped around the plurality of creative design, open compositional practices, and the reality of versioning?

We are moving from an era of scholarship based on the individual author of the "great book" to an era of scholarship based on the collaborative authoring possibilities of the "great project." Because we are in the midst of a transformation in the materiality of information and in the media technologies of communication, things that were once considered "mere" support systems, transmission media, and conveyance devices are now fundamentally implicated in any meaning-making process. Great books do not simply "contain" great ideas but are part of a techno-social apparatus of inscription and alphabetization. One may study the history of the page as a spatial unit of order; the material history of paper, ink, printing blocks, and printing presses; and the navigation of the object by the intervention of the human body. Print culture's centuries of stability undermined humanists' ability to "see" the materiality of their practices: the book became a transparent medium. Digital humanists, on the other hand, foreground the deeply recursive ways in which meaning and interpretation are bound up with materiality, media, and embodied navigation. This is why we stress that authorship is design and design is authorship.

Within the Digital Humanities, knowledge platforms cannot be simply left to editors, technicians, publishers, and librarians, as if the physical and virtual arrangements of information as argument through multimedial constellations are somehow not the domain of humanities scholars. In the print model, scholars typically "handed off" the content of their manuscripts to publishers who did the layout, design, editing, printing, and dissemination of the work. Now, however, these tasks have moved to the forefront of the Digital Humanities precisely because choices of interface, interactivity, database design, mark-up, navigation, access, dissemination, and archiving are all part of how arguments are staged in the digital world. These choices are evident, for example, in the projects published in *Vectors*, a multi-modal, multimedia humanities journal in which each "article" is a project that explores the complex interrelation between form and content, underscoring the "immersive and experiential dimensions of emerging scholarly vernaculars across media platforms." Scholars work closely with designers, technologists, and the *Vectors* editorial team to develop appropriate interfaces, database schemas, navigation features, and content types that, altogether, instantiate an argument. While preserving the authority of peer review, the publication platform not only foregrounds collaborative authorship, but also public feedback through threaded discussion forums and annotation features.

The challenge for the Digital Humanities is to develop the evaluative metrics for legitimizing and credentialing this kind of scholarship since it places a high demand on readers to participate, navigate, explore, interact, and often contribute. A project published in *Vectors* may have multiple "authors," each of whom contributed to the argument: interface designers who created a Flash front-end, database designers who created a MySQL/PHP back-end, programmers who wrote the code to interact with the database and parse queries, academic scholars who populated the database and designed an interactive architecture for navigating the argument, GIS specialists who formatted and processed the data, modelers who created navigable 3-D models of physical environments, server administrators who oversee the appropriate operating software to ensure that the project remains functional, and so forth. Much like science articles with multiple co-authors, it is already common for Digital Humanities publications to list a series of authors on a project, and this is expanded exponentially when we are talking about the development of a platform. The singularity of the "I-subject" has been transformed into the collaborative authorship of a "we-subject."

Collaboration as Creation

These recalibrations are informed by and contribute to what we have been calling the social life of the Digital Humanities. Even where scholarship still "looks" like a book written by a single author, we are now witnessing the first wave of creative destruction of long-held truisms, behaviors, and practices in the academy. Some scholars and artists have published versions of their books online using paragraph-by-paragraph blogging software or other collaborative annotation and commenting engines. Not only does this repudiate the notion of intellectual property as something locked up by copyright and exclusive licensing agreements, it also allows the authors to receive immediate feedback by hundreds of self-selecting peer reviewers—before the book is sent, by a university press, to a couple of scholarly authorities in their field. Crowd-sourced evaluations of scholarly arguments, not to mention crowd-sourced production models for generating and editing scholarly content, are transforming both the authorship function and conventional knowledge platforms: A book is not simply "finished" and "published," but is now part of a much more dynamic, iterative, and dialogical environment that is predicated on versioning, crowd-sourced models of engagement and peer review, and open-source knowledge and publication platforms.

This is nowhere more apparent than with Wikipedia, a revolutionary knowledge production and editing platform. While Wikipedia was dismissed by many within the academy as amateurish, unreliable and lacking in scholarly rigor (especially in its early years), we suggest that it is a model for rethinking collaborative research and the dissemination of knowledge in the Digital Humanities and throughout academia. Wikipedia represents a truly innovative, global, multilingual, collaborative knowledge-generating community and platform for authoring, editing, distributing, and versioning knowledge. To date, it has nearly 4 million articles, more than 450 million edits, more than 15 million registered users, and articles in scores of languages. This is a massive achievement for the first decade of work. Wikipedia represents a dynamic, flexible, and open-ended network for knowledge creation and distribution that underscores process, collaboration, access, interactivity, and creativity, with an editing model and versioning system that documents every contingent decision made by every contributing author. At this moment in its short life, Wikipedia is already the most comprehensive, representative, and pervasive participatory platform for knowledge production ever created by humankind. That is worth some pause and reflection.

It is striking that Wikipedia was not invented at a university, and though one of its founders has a humanities Ph.D., it operates outside the academy. Why might this be? Perhaps because the humanities—in broad strokes—remain fixated on discrete publications by individual scholars, primarily in conversation with others like themselves, working in single media forms. It is one thing to create new knowledge within the theoretical, methodological, material, and disciplinary paradigms of a field; it is something quite different to imagine a new knowledge platform, a new way of designing knowledge and engaging broad communities in knowledge creation. What this means in practice is that as we shape our platforms, tools, and technologies, our platforms, tools, and technologies shape us. These mutually reinforcing systems form the social life of the Digital Humanities. They are mutually co-constitutive and profoundly recursive in ways that are generating new notions of what it means to be a human being as a subject that knows, as a creator of knowledge, and as an object of study.

Publishing as a Social Act

"To publish" is to make something public, to place it within a sphere for broad scrutiny, critical engagement, and community debate. Traditionally, publishing meant finding a journal or press in order to make academic treatises, arguments, and the results of research public—but this "public" was in reality primarily or even exclusively readers initiated in and defined by the discursive conventions of a given field. Today, almost anyone can publish (in the sense of "make public") anything. As noted earlier, the bar to entry for starting a blog, tweeting messages, posting photographs or videos, hosting a website, or commenting on other people's blogs, messages, postings, and websites is extremely low. It's not uncommon for a video that has "gone viral" to amass tens of millions of views across the globe within days or even hours. Clearly, we are witnessing yet another contraction of time and space, as information is radically decoupled from the specific identity of the creator.

For scholarship to engage with this contraction, let alone the unbinding of argument from author, raises serious questions for the humanities, which has, traditionally, considered a "proper" publication to be a peer-reviewed, vetted argument that cites and speaks to the conventions of a particular discourse and represents the views of an author who has gained authority by having passed a series of "tests" that credential the author to speak in legitimated utterances. Authors are generally affiliated with an institution which grants authority to their utterances by virtue of various rules of inclusion and exclusion (i.e., the tenure and promotion system, the imprimatur of book and journal publishing, the grant and foundation support industry, and so forth). The places where works are published, such as journals and university press books, have established themselves as authoritative sources of knowledge by virtue of strict mechanisms for peer review, scholarly vetting, and institutional reputation which has been built up over decades, sometimes even centuries.

What happens when anyone can speak and publish? What happens when knowledge credentialing is no longer controlled solely by institutions of higher learning?

These are serious questions confronting the institutions that function within and maintain the social life of the Digital Humanities.

Transforming Publishing and Access

Scholars in the humanities have become used to social norms of knowledge formation and dissemination. Nothing seems more natural to this social structure than the idea that scholars write manuscripts, that publishers produce them as books, and that libraries aggregate them as collections and provide access and other services for reading and research. The inherited norm is that publishers commission authors or acquire intellectual property they deem worthy of making public. They review the manuscript's content and argument, and check for originality and legitimacy. A scholarly publication would elicit peer reviews. Fact-checking, line-editing, permissions for illustrations, layout, design, printing, and advertising all require a set of skills and professional expertise. Similarly, the tasks of librarians are specialized. Institutions differ, patrons have a host of varied profiles, and the needs of any particular library are specific to its setting and the services it provides, but traditionally these tasks have included acquisitions, cataloging, preservation, conservation, public services, outreach, and access. Digital publishing models, however, are challenging these long-standing roles and institutional boundaries.

A certain tension exists in the current environment as libraries and publishers confront a changing landscape, but it is important to state certain obligations that remain vital to humanistic inquiry no matter how technologies affect social constructions. The recent tight budgets for scholarly presses have pushed for reconsideration of the business models developed in the print environment. Formats are changing, but peer-review—which can now be extended even to the public sphere—remains crucial. Timelines and life cycles of information are shifting, but the need for reliable references remains; perhaps it is more urgent than ever. Licensing agreements and expectations about long-term access must be addressed as must the recognition that a new business model has to emerge that takes seriously issues involving the evolution of intellectual property, open-source culture, copyright protections, and what has been referred to as the challenge of "copyleft" considerations. Print-based understandings of concepts such as first sale—buying a copy of a book grants the right to pass the book on—are problematic in a digital environment in which a copy of a text or work can be easily replicated and distributed. What are the rights of authors and of presses? How does society balance these rights against the needs of readers, scholars, libraries, and the broader public?

Emerging Fellowships of Discourse

The university has long shared the tasks of knowledge production, curation, stewardship, and storage with other cultural institutions such as laboratories, publishers, libraries, museums, and commercial producers. But the university legitimates knowledge in a privileged way, supervising rules of admission to and control over discourse. Not just anyone can speak with authority; one must first be sanctioned through lengthy and decidedly hierarchical processes, and the knowledge that is transmitted is primarily circulated within relatively closed communities of knowers, "fellowships of discourse" as Michel Foucault termed them. Statements are repeated and circulated through various disciplinary and institutional forms of control that legitimize what a "true statement" is within a given discipline. Before a statement can even be admitted to debate, it must first be, as Foucault argued repeatedly, "within the true." For an idea to fall "within the true," it must not only cite the normative truths of a given discipline, but it must fall "within the true" in terms of its methodology, medium, and mode of dissemination. Research articles can't be Wiki entries; book monographs can't be exhibitions curated in virtual worlds; seminars can't be held in gaming environments. Or can they?

What is at stake is a question of legitimation. **Who can create knowledge, who monitors it, who authorizes it, who disseminates it, whom does it influence and to what effect?** Legitimation is always, of course, connected with power, whether the power of a legal system, a government, a military, a board of directors, an information management system, the tenure and promotion system, the book publishing industry, a professional group, or any oversight agency. Not only are discursive statements legitimized by the standards established by the practitioners and history of a given discipline, but so are the media in which such utterances are formulated, articulated, and disseminated.

The authorship function in the Digital Humanities is more collaborative, involving designers, coders, information architects, and server administrators, not to mention scholars from adjacent and nonadjacent disciplinary fields. And the notion of "the work" is significantly more porous and process-oriented, requiring a very different set of criteria to evaluate its merits. In the past, hermeneutic analysis sufficed because peer reviewers privileged the "insides" of a text: that is, they privileged what was said, how it was substantiated, and what was argued. An original argument pushed the boundaries of a given field forward but still operated within the theoretical, disciplinary, and media-specific paradigms of knowledge in that field. The medium that conveyed the argument was rendered transparent and neutral as in Beatrice Warde's long-cited image of the well-designed book as a "crystal goblet."

Digital Humanities denaturalizes print, awakening us to the importance of what N. Katherine Hayles calls "media-specific analysis" in order to focus attention on the technologies of inscription, the material support, the systems of writing down ("*Aufschreibesysteme*," as Friedrich Kittler puts it), the modes of navigation (whether turning pages or waving your hand), and the forms of authorship and creativity (not only of content but also of typography, page layout, and design). In this watershed moment, awareness of media-specificity is nearly inescapable and carries implications for the social life of these media as well.

Shaping New Norms

With the rise of new authoring platforms and collaborative environments, "supporting" apparatuses have been exposed as anything but transparent and neutral, as they not only determine modes of interaction and navigation but also condition and guide the production of meaning. Publication is not an endpoint or culmination of research, but is something significantly more process-oriented, indeterminate, experimental, and even experiential. Therefore, a whole new set of evaluative questions needs to be asked. We might take the following as new normative questions for evaluating humanities scholarship in general—that is to say, not just Digital Humanities scholarship:

How does the work present and advance an original argument that is bound up with and a function of the materiality and medium in which the argument is presented? In other words, what does materiality and media mean for the instantiation of the argument?

Who are the authors of the work and how are their contributions articulated and credited?

How does the design of the interface, the data structures, and the database convey meaning and function as part of the argument? How does a reader interact with the work, and how do the authors expose the rhetorical elements of their interface, data structures, and database?

Is the mode of navigation and kinetic signposting appropriate for the argument?

How complete is the bibliographic apparatus of the work and how do readers access both the sources cited and the data presented?

Can the work be deployed and enhanced by putting it in new contexts or in new digital environments with other projects?

Is the work extensible and iterative? That is to say, can it continue to grow as more research is done either by the author or other people?

How can the participatory dimension of the work be characterized? In other words, does the argument demand greater participation than page-turning or mouse clicks?

Does the scholarship support federated (non-silo based) approaches to scholarly publishing?

Above all, how does the work embody standards of traditional scholarship that can inspire a broad community with its insights?

These kinds of questions interject a different set of evaluative metrics into humanities scholarship while raising the bar for digital work. We are still at the very earliest stages of understanding and legitimating these emergent knowledge formations. We do not want to lose sight of the core values by which scholarship is judged, and we also want to be sure we can answer skeptics ready to assert that the Digital Humanities is all technique and lacks content.

Such a balanced approach not only underscores a fundamental rethinking of *how* knowledge gets designed and created, but also a fundamental rethinking of *what* knowledge looks and sounds like, *who* gets to create and interact with knowledge, *when* it is made and recognized, *how* it gets authorized and evaluated, and *how* it is made accessible to a significantly broader (and potentially global) audience. This is why we must discuss the social life of the Digital Humanities holistically, rather than follow a piecemeal, instrumentalist approach. In the 21ST century, long-established institutions like universities and their presses have the potential to generate, legitimate, and disseminate knowledge in radically new ways, on a scale never before realized, involving technologies and communities that rarely (if ever) were engaged in a global knowledge-creation enterprise. We are just starting to understand and leverage that potential, and the question is how to sustain (and not short-circuit) this critical process of experimentation and risk-taking.

Decolonizing Knowledge

The ways in which we have been discussing the social life of the Digital Humanities have privileged technology's transformative impact upon scholarship. But there is a reciprocity that is less visible but equally important: The principles of humanist thinking, humanist creativity, and humanist critique have much to offer to computational methods. Humanistic design of digital environments can challenge and even undo the normative assumptions that encode ideological assumptions in operational features. Efficiency and transparency have been bywords of interface design. Yet digital humanists can imagine means to model the complex conditions of interpretation so that we come to a fundamentally different idea or demonstration of the ways engagement with the cognitive processes of reading, viewing, and navigating make meaning. The participatory environment of the creation of cultural materials calls for analysis and display of the co-dependent relation between communities of thought and their expression. We have yet to engage seriously with modeling environments that support cultural difference, rather than register it, often in static and even monolithic ways, on standard platforms developed by dominant industry players. **If the platforms set the terms of cultural production, then whose worldviews and ideologies will they embody and structure into the creation of knowledge?** Might we envision alternatives, for instance, to mapping the beliefs of indigenous peoples onto a Sloan Digital Sky Survey, and instead remake the presentation of the sky in the form of such beliefs? It is not that such interfaces and affordances "change" the sky so much that our appreciation of how people "see" the heavens becomes both deeper and broader. The decolonization of knowledge in the most profound sense will arrive only when we enable people to express their otherness, their difference, and their selves, through truly social and participatory forms of cultural creation.

If the organization of and navigation through information are statically structured, we move through massive amounts of material but do not change the ontologies, the very ways of knowing, that govern storage, access, and display. Humanistic interfaces are social as well as technological, and so will mutate and change, remaking the order of the knowledge field in response to modes of engagement, interpretive gestures, and linguistic and cultural differences. We have yet to fully examine and expose the historical dimensions of classification systems, epistemologies, and knowledge representations in ways that model and present their incommensurabilities across cultures, historical periods, and individual understandings. We must interrogate the spaces for the production of what gets to count as knowledge at a given moment, the modalities for the production and ordering of discourse, and the conditions of possibility for the configuration of knowledge into systems, classification schemata, representations, and ordering principles.

Bringing these fundamental features of humanistic inquiry into the digital environment is also essential work for the Digital Humanities. Building and using tools that are rooted in traditional humanities concerns—subjectivity, ambiguity, observer-dependent variables in the production of knowledge, contingency—will allow us to model knowledge and creative work both ontologically and socially. The next generation of Digital Humanities work will make a contribution to theory only if it can show how to think *in* digital methods, not just *with* digital tools. Indigenous, local, independent, and truly alternative humanities platforms are still only speculative concepts, latent, perhaps on the verge of emergence. The excitement lies in envisioning these possibilities and imagining how to shape future knowledge production along lines as yet unthought, unmapped, and unsaid. We need to take seriously the conviction that the humanities have their own methods—not based in calculation, automation, or statistical probability, but in ambiguity, interpretation, and in embodied and situated models of knowledge and knowing.

Revitalizing the Cultural Record

By conceiving of scholarship in ways that significantly involve community partners, cultural institutions, the private sector, nonprofits, and/or government agencies, the Digital Humanities expands both the notion of scholarship and the public sphere in order to create new sites and nodes of engagement. With such an expanded definition of scholarship, digital humanists are able to place questions of social justice and civic engagement, for example, front-and-center. They are able to revitalize the cultural record in ways that involve citizens in the academic enterprise and bring the academy into the expanded public sphere. The result is a form of scholarship that is, by definition, applied: It applies the knowledge and methods of the humanities to pose new questions, to design new possibilities, and to create citizen-scholars who value the complexity, ambiguity, and differences that comprise our cultural record as a species.

By foregrounding the values of the humanities, such projects create an environment in which silenced voices, cultural differences, linguistic multiplicity, and historical perspectives vitally inform and expand the notion of "public" and the "public sphere." Documentary projects that integrate social media offer compelling examples of how technologies like Twitter can be used to give voice to people who are otherwise silenced. Crucial political events since the advent of social networking have shown how highly localized and accurate accounts of what was happening on the ground can be assembled using a combination of random and trusted informants, including everything from simultaneous postings to live feeds and messaging platforms, often with links to audio files and other media reports that help the world "see" and "hear" what is going on in real time. In effect, the digital portal becomes a global public sphere linked to precisely located events that, in turn, become part of a Web archive and living memorial.

While the "role" of social media has been feverishly debated in fomenting, planning, and sustaining revolutions since Twitter was first hailed as a revolutionary technology in Moldova in 2009 and YouTube became a living archive for election protests in Tehran during the summer of that same year, it seems incontestable that "something" is happening to media that is changing the way in which events unfold. If nothing else, there is a massive contraction and alignment of the event (an embodied and location-specific phenomenon), the representation of the event (through tweets, cellphone video and photographs, and so forth), and the dissemination of the event (through Web-based social networks and information channels). The result is a significantly more adaptable, amorphous, global, but also ephemeral public sphere, one which may, for example, be constituted in distributed locations simultaneously.

Publics and Counterpublics

While physical embodiment becomes simultaneously less and more important in constituting a public, it is also worth remembering that media and communication technologies have always played a fundamental role in creating what is understood as the public sphere. Jürgen Habermas' acclaimed study of the structural transformation of the public sphere showed it to be an invention of bourgeois society in the late 17TH and early 18TH centuries that came about through the rise of newspapers and novels as well as through new forms of sociability that encouraged discussion and debate. Print technologies and the spread of literacy were critical for the formation of "the public" and the rise of the modern nation-state, with the former specifically arrayed against the state as a locus of authority. Kant considered the "public use" of reason to be that of "a scholar before the entire public of the reading world," a definition that also betrayed the conspicuous limits of that term: Kant's public was constituted by literate men, who became literate because of their belonging to a particular socioeconomic stratum.

"Counterpublics" emerged as a parallel phenomenon, constituted by intellectuals, some of them outstanding women authors who organized salons in their homes—in tension with, often against, but still connected to, public discourse. In this regard, notions of "the public" and "the counterpublic" are exclusive and often even elite formations precisely because the admittance of members to discourse is socially and economically determined. More recently, attention has been paid to the discourses of the "subaltern," those whose class, race, and gender positions situate them fundamentally outside any dialectic of "public" and "counterpublic," creating dialogues that are barely recognizable as "public speech" because they do not stem from "within the true," as Foucault put it.

Perhaps, then, the utopian impulse of the Digital Humanities can be characterized as a modality of radically opening discourse to participation for everyone. **What if there were participation without condition?** What if utterances were neither admitted nor denied based on gender, sex, race, ethnicity, language, location, nationality, class, or access to technology? We are not saying that these facticities do not matter or cease to matter in the digital world; instead, we are saying that the utopian element of the Digital Humanities is to at least posit, if not fully enable, a future in which participation is possible for everyone, anywhere, anytime. It would be *as if* it were possible to bring about a public sphere in which no one was excluded. This is a core human value of the Digital Humanities.

Electronic Presses and Ubiquitous Libraries

For many centuries, university presses have played a crucial role in establishing the currency of the humanistic profession in formats and practices. Monographs, edited collections, critical editions, and scholarly journals are the basic elements of research and professional development. Careers are made on the basis of vetted and peer-reviewed literature in the form of essays and book manuscripts. The presses have the expertise to create marketing plans, assess audience, and develop distribution networks with libraries and scholars. Acquisitions editors keep tabs on their fields with expert attention—attending conferences, tracking the intellectual development of disciplines, and helping shape the discourse in any particular field by the work they recognize through publication. This bedrock expertise combined with a professional commitment to disciplines and discourses must continue to be supported by salaried jobs and institutional frameworks even as the social and economic conditions of academic publishing change in the digital era.

As the social life of the Digital Humanities evolves, many university research libraries are also reconsidering their charge. Can they continue to afford to collect serial work that essentially buys the right to distribute intellectual research that has been created by faculty on their own campus? This is the real problem of skyrocketing scholarly journal prices. And what happens to licensed material if a service provider or company goes out of business and the link to published work disappears? Here we confront the issues of bit rot, technological obsolescence, and the risks of investing in emergent technologies. Many considerations enter into the mix. Are copies stored on local servers in the library? Or are links to a repository the only means of accessing intellectual work? Many of the thorniest problems are social rather than technological. Putting knowledge in protected silos, areas in which scholarship is only available to a limited community of academic professionals, can hugely benefit those select scholars. But such lockdowns go against the impulse to bring the best cutting-edge work before the broadest possible audience.

Add to this mix the problem of finding a recognized and visible portal for exposing new digitally published research. In particular, if the granularity of contributions changes, so that annotations, code, data sets, or large-scale processing are considered units of argument, then where and how will these be recognized and acknowledged? If posting becomes equivalent to publishing (on a blog or a social media site, via a live feed or other as-yet-to-be-imagined platform), then our definitions of scholarly publishing and our traditional obligation to preserve, catalog, and provide access and reference to these pieces will push us toward radically new understandings of the roles of both libraries and publishers. Distribution mechanisms will need to evolve in ways that recognize the productive distinction between popular work and more specialized scholarship, and address the complex set of issues that will continue to emerge around intellectual property, licensing and use, peer-review, and the role of professionals in publishing, preserving, and providing access to scholarship. The challenge of maintaining platforms, as well as works, will only complicate matters further, as the iterative versions of software and hardware for access and display, driven by market forces and industry agendas, compete with the longevity and stability that print forms have accustomed us to over the centuries.

To be sure, the need to avoid redundancy and optimize resources will drive part of the reconfiguration of publishing in the digital realm. But the theoretical issues remain: What is a publication? **Who will undertake the making-public of arguments, research projects, repositories, archives, and other materials of the human record, its creative expression and interpretation?** New concepts are already transforming notions of publishing, publicity, and the public. The digital turn in scholarship is bringing into view genres undreamt of in earlier media. As it does this, libraries and publishers will forge alliances that distribute old tasks along new lines as they take on novel responsibilities and forms of engagement unforeseen in an analog world.

The Care and Feeding of Hedgefoxes

Digital Humanities has many goals. Some involve research; some focus on outreach to broader publics; some are pedagogical in nature. One of the fundamental questions confronting Digital Humanities is what kind of student will its methods produce? If the academy and society support Digital Humanities, what kinds of students will they train and how will these students shape the world? An alternate method is to imagine the kind of students one would like to see, and then work backward to envision the educational environments most conducive to producing such a cohort. This kind of hypothetical persona-building allows us to reinvigorate all-but-exhausted discussions about the broader implications of a liberal arts education, and ties these issues back to the discussion of Digital Humanities as forming a core curriculum. The kind of student universities train leads to the questions of what sort of citizens they can become, how they will function as autonomous individuals, and how they will integrate themselves into society.

To think through these questions, it is worth reaching back. Two-and-a-half millennia ago, the Greek poet Archilochus broke the world of knowledge into two camps, represented by two different types: "the fox knows many things, but the hedgehog knows one big thing." Half a century ago, Isaiah Berlin reworked this metaphor to divide thinkers "between those... who relate everything to a single central vision, one system less or more coherent or articulate... [and] those who pursue many ends, often unrelated and even contradictory, connected, if at all, only in some de facto way." Berlin made no claims for the superiority of the ways of either the fox or the hedgehog, and devoted an essay to the productive conflict Leo Tolstoy generated as a fox who thought he was a hedgehog.

We are in an era far different from the Greek poet's, the Russian novelist's, and the English don's. There can be little doubt that the technologies that give rise to the Digital Humanities push us—scholars, students, and citizens alike—into the fox family. The nature of discourse and debate in networks, the reality of study in multimedia environments, and the inexorable splintering of attention that multiple windows and channels afford lead to pursuing "many ends." This tendency toward multi-tasking and shortened attention has a multitude of detractors, of course, as well as the usual contrarian supporters of the "everything bad for you is good for you" variety. But the Digital Humanities can confront this reality on the ground (and in the ether) without either nostalgia for a reader's paradise that never was or the kind of hype over technology that we expect from industry. The Digital Humanities has methodologies that can harness the habits and possibilities of the minds of a networked generation to create better and more inquisitive foxes.

Yet what of the hedgehog? It is precisely the hedgehog's tenacity, its willingness to spend months, years, and decades in pursuit of a "single central vision" that ties it to the practice of the humanities. There are fewer opportunities for the long haul and the deep dig in a society that embraces the business quarter, instant access, and machine time. The traditions of the humanities, on the other hand, embrace the durational, accepting that some studies will take years to complete, that certain ideas, needless to say conclusions, demand lengthy gestation. The multivolume study, the life devoted to a specific slice of a discipline, these are the hallmarks of the humanities, and the Digital Humanities would be foolish indeed to abandon its inner hedgehog.

How can the Digital Humanities keep the ways of the hedgehog alive in the era of the fox's ascendance? How do we inject deep digs into the free-ranging ways of networked scholarship? The hedgehog's great depth is inspiring for its rigor; the fox's curiosity is astonishing in its energy. It is not an either/or situation: the goal is hybridization, the creation of hedgefoxes, capable of ranging wide, but also of going deep. Making the move from creating, appreciating, and interpreting the hedgefox aesthetic to responsible, 21ST century citizenship requires that students of Digital Humanities see social networks as having both pro- and anti-social agendas, that they develop political literacies, and that they harness the collaborative energy of their academic experiences and apply them to the broader culture.

4. PROVOCATIONS

THE ERA OF DIGITAL HUMANITIES HAS JUST BEGUN, BUT IT MAY BE COMING TO AN END.

TWO DECADES AGO, WORKING WITH DIGITAL DOCUMENTS WAS THE EXCEPTION. TODAY IT IS THE NORM, THE "NATURAL" ENVIRONMENT FOR CARRYING OUT RESEARCH, TEACHING, AND READING. WIRELESS NETWORKS HAVE CONSIGNED THE OFF-THE-GRID, OFF-LINE CLASSROOM TO THE DUSTBIN OF HISTORY. IF THE NOVELTY OF DIGITAL HUMANITIES WORK HAS ALREADY BEEN ABSORBED INTO DAY-TO-DAY BUSINESS, THEN WHAT CLAIMS CAN IT MAKE FOR INTRODUCING NEW INSIGHTS OR METHODS INTO RESEARCH FOR THE BROADER FIELDS OF THE HUMANITIES? THE PRACTICE OF DIGITAL HUMANITIES CANNOT BE REDUCED TO "DOING THE HUMANITIES DIGITALLY"; BOTH CRITICALITY AND EXPERIMENTATION MUST SHAPE ITS FUTURE DEVELOPMENT.

Decades of work were involved in building digital repositories and establishing conventions for access and use, not to mention in developing the communication, presentation, and publication tools upon which humanists rely for information-sharing and dissemination. Each of these undertakings represents an act of interpretation. Every migration from analog to digital is a translation that stages *a certain experience* of artifacts encountered online. Advanced Web technologies add a social dimension into online experiences. Other developments will follow and create new operational possibilities as well as new constraints, absences, and blind spots.

When new norms establish themselves, when new procedures and techniques become naturalized, assumptions can become invisible. The pressure to reflect critically, to innovate and alter consolidated practices, can subside. Digital Humanities is still in its infancy. But its ability to serve as a driver of innovation could become threatened as "doing the humanities digitally" becomes business as usual. Could the era of Digital Humanities come to an abrupt end? Perhaps it could. Or it might give rise, phoenix-like, to a new spirit of experimentation if "doing the humanities digitally" is accompanied by the same spirit of innovation that fueled the first generation of digital work. That will require thinking creatively and experimentally.

Today's humanists work fluidly across the digital/analog divide. Venerable binarisms have begun to blur into a continuous workflow. Though even digital natives recognize the difference between a manuscript held in one's hands and one viewed on the screen, the space of engagement created by working across media tends to collapse differences and create an illusion of frictionless exchangeability. We sit with books in front of us, typing notes into files, watching videos on other screens, and tracking references through search engines on windows open on our laptops. Are we all digital humanists? No. Are we carrying out the work of the humanities digitally? Routinely so.

But the new routines that structure this world of practice have the potential to become just as sedimented and automatic as those of the print era, and when they do, they sound the death knell for Digital Humanities as a practice that is both critical and experimental. How will Digital Humanities continue to provide ways of thinking differently about the methods and objects of study that constitute knowledge?

AS DIGITAL TOOLS BECOME NATURALIZED, THE DIGITAL HUMANITIES WILL STRUGGLE TO RETAIN ITS CRITICAL, EXPERIMENTAL CHARACTER.

Maintaining criticality and experimentation means challenging received traditions, even—perhaps, especially—those that defined the first generations of Digital Humanities work. Innovative forms of public engagement, new publishing models, imaginative ways of structuring humanistic work, and new units of argument will come to take their place beside the pioneering projects of the first generation. This means embracing new skill-sets that are not necessarily associated with traditional humanistic training: design, programming, statistical analysis, data visualization, and data-mining. And this means developing new humanities-specific ways of modeling knowledge and interpretation in the digital domain. It means showing that interpretation is rethought through the encounter with computational methods and that computational methods are rethought through the encounter with humanistic modes of knowing.

THE HUMANITIES NEED TO ESTABLISH DISCIPLINE-SPECIFIC AGENDAS FOR COMPUTATIONAL PRACTICE.

But, as they do so, the toolkits they employ and the topics they tackle may become as attached or detached from contemporary societal discourse as are today's languages of critical theory and cultural critique. Humanists have begun to use programming languages. But they have yet to create programming languages of their own: languages that can come to grips with, for example, such fundamental attributes of cultural communication and traditional objects of humanistic scrutiny as nuance, inflection, undertone, irony, and ambivalence.

Is computational work fated to remain locked in the realm of quantifiable and repeatable phenomena? How might one model the complex dynamics of interpretation or the processes by means of which reading, viewing, and playing generate cultural meanings within a given community or tradition? How might techniques like probabilistic modeling, interpretive mapping, subjective visualizations, and self-customizing navigation alter our experience of the digital realm and the character of the Web as a public domain?

The navigation of information remains structured in a static manner despite the fact that it is experienced dynamically. Users rarely engage with or alter the ontologies that govern the storage and display of material: Their responses don't shape the information architecture, only the contents it will serve up. Yet the study of the human cultural record demonstrates that, far from innate, such architectures are built upon classification systems and knowledge representations that differ across cultures, historical periods, and even the worldviews of different individuals within a single culture.

This raises the question of how humanistic ways of doing and thinking might be brought to bear in the domains of knowledge retrieval, curation, and use. Imagine, for instance, a Heraclitean interface, a hybrid of the very old and the very new, founded on notions of flux and the non-self-identical nature of experience. Such an interface might mutate and change, shifting ontologies on the fly, remaking the order of the knowledge field in response to a user's queries and reactions to the results. It might well work like a dream for performing certain research tasks (like studying dreams) and like a nightmare for others (like counting hedgehogs or proving a point).

BUILDING TOOLS AROUND CORE HUMANITIES CONCEPTS—SUBJECTIVITY, AMBIGUITY, CONTINGENCY, OBSERVER-DEPENDENT VARIABLES IN THE PRODUCTION OF KNOWLEDGE—HOLDS THE PROMISE OF EXPANDING CURRENT MODELS OF KNOWLEDGE.

AS SUCH, THE NEXT GENERATION OF DIGITAL EXPERIMENTERS COULD CONTRIBUTE TO HUMANITIES THEORY BY FORGING TOOLS THAT QUITE LITERALLY EMBODY HUMANITIES-CENTERED VIEWS REGARDING THE WORLD.

Tools are not just tools. They are cognitive interfaces that presuppose forms of mental and physical discipline and organization. By scripting an action, they produce and transmit knowledge, and, in turn, model a world. In the case of an indigenous humanities digital toolkit, the world in question might not be the same as that envisioned by the server farm that monitors inventory levels at Walmart warehouses, or the ones envisioned by the systems that track air traffic over the Pacific or match banner ads to the content of Gmail accounts. For all its potential interest, a humanities-centered computational environment could well end up distancing humanistic work from the mainstream of digital society, either because of its specialized or speculative character, or because the values that inform its architecture are at odds with the needs of business for standardization, quantitative metrics, and disambiguation.

AS HUMANS AND DATA MACHINES BECOME EQUAL PARTNERS IN CULTURAL PRACTICE, SOCIAL EXPERIENCE, AND HUMANISTIC RESEARCH, THE HUMANITIES MAY NO LONGER LOOK LIKE "THE HUMANITIES."

THE SCALES AND REGISTERS OF WHAT COUNTS OR IS VALUED AS HUMAN EXPERIENCE AND, THEREFORE, THE OBJECTS OF HUMANISTIC INQUIRY, WILL FIND THEMSELVES ALTERED.

The cognitive horizons of digital researchers are already being deeply altered by the ability of data machines to zoom back and forth between grand sweeping views of masses of texts, data, and images and the microscopic particulars of single documents or objects. Trust in computers' capacities for aggregation, synthesis, and even selectivity is sure to grow over the coming century. Visions of machine agency and emerging sentience reek of science fiction fantasies, but unintended consequences may well be in our future.

Will we read the machine's analyses and summaries of marked texts, structured data, and natural language processing and feel we are in conversation with an adept partner, whom we will be tempted to imagine as a natural extension of our own cognitive capacities? Our partner will not only provide information, but also parry our every query with alternatives and suggestions for reflection. Perhaps we will become ever-more seduced by the macro and micro ends of the perceptual spectrum, by very big and very small data. We may become ever-more inclined to neglect the in-between realm within which most of human experience has unfolded over the millennia. In a kingdom in which zoomability rules, linear reading may seem akin to a horse-and-buggy ride. On the other hand, these technophilic projections may themselves come to seem dated, quaint relics of an era before reading in an expanded field was the norm.

In reality, the machine may provide conceptual frames and filters that provide access to, process, and shape the historical record. Analysis of materials concerning the relative significance of the seas during 10 centuries of human history, for instance, will generate ecological, biological, chemical, political, literary, artistic, or geological filters. Each framework creates a different synthesis with pointers toward higher and lower levels of aggregation as well as specific documents, materials, views, models, or other evidence on which the synthesis was compiled.

Complex adaptive systems theory suggests that our ways of understanding knowledge production in social systems could expand (but always at a cost, given the finitude of available cognitive resources). We might come to recognize that what occurs in communities and through communications networks takes a shape that emerges from the patterns of collective activity, rather than merely being the aggregate of individuals' actions. At this higher level, networked systems of exchange produce a thinking effect, simulacral or real, a sense of knowledge produced at the system level. What, for instance, is the sum of the phone calls on a given evening from one point to another when these are not seen as an aggregate of individual conversations, but as the pattern of the network itself? How do such patterns influence the activities and perceptions of individuals through the coercive force of normativity? Where and how are ideas stored in the *noösphere*? What is the medium of collective thought and how might it be realized and understood, visualized, analyzed, grasped?

Only a small segment of the humanities community needs be excited by the design of projects and protocols to address questions of these kinds. Most will be happy to be users of digital domains. Phrases like "distant reading," "content modeling," or "knowledge representation" will become just as familiar a part of our vocabulary as the terms "social media" and "networking" have.

A TENSION EXISTS IN THE CONTEMPORARY ERA OF THE DIGITAL HUMANITIES, WITH ONE WING OF THE HUMANITIES EMBRACING QUANTITATIVE METHODS, THE OTHER CONTINUING TO INSIST UPON ITS ROOTS IN QUALITATIVE ANALYSIS.

THE QUANTITATIVE WING BECOMES INTEGRATED INTO THE SOCIAL SCIENCES. THE OTHER FIGHTS TO DEFEND ITS AUTONOMY AND CRITICAL STANCE.

PARTNERSHIPS AND PARTISANSHIP LIE AHEAD.

The prodigious ability of computers to work with large data sets, be they of inventories in automobile showrooms or research libraries or collections of ancient papyrus fragments, leads to a bifurcation within the Digital Humanities community that can be traced back to its beginnings.

The proponents of big data analysis seek to marshal these powers to undertake tasks that exceed the scale of mainstream humanistic inquiry, arguing for a close alignment with the quantitative social sciences. They look beyond the qualitative or interpretive preoccupations of traditional humanistic inquiry in favor of standard social science statistical methods, preferring a macro scale of analysis and description in order to examine such questions as the dissemination patterns of cultural forms, the shape of literary marketplaces and reading habits, and shifts in the physical and other external attributes of aesthetic objects.

The primacy of big data research is rejected by digital humanists who consider such methods epistemologically naive and their results generally trivial, self-evident, or flawed. They argue that the tools of the empirical sciences—statistical graphs and visualizations, grids and charts, maps and tables—carry conviction because they assume information is observer-independent and rooted in certainty. And while they are willing to admit that such techniques have played and will continue to play important roles in expanding the compass of Digital Humanities research, they contend that such tools are ill-equipped to capture the complexities of novelistic constructions of character or to trace the day-to-day, document-to-document shifts in tone found in a statesman's archive that translate into policy shifts and alter world affairs. At stake for them in tracking this elusive universe of signs is much more than the mismatch between qualitative judgments and the quantitative strictures imposed by analytic tools or graphical expressions borrowed from the social sciences. What is at stake is the humanities' unique commitment to wrestle with uncertainty, ambiguity, and complexity; to model incommensurate temporalities and ontologies; to explore not just geographies but psychogeographies and the dark recesses of the self; to attend to non-repeatable and nonstandard phenomena. Such forms of attention are freighted with special meaning inasmuch as they closely correlate with the critical function that the modern humanities disciplines have performed in contemporary society: their championing of difference and the non-normative, their assault on sedimented social behaviors and norms, their ability to defamiliarize and historicize social institutions.

The battle rages on with both sides freely encroaching on one another's turf, with an itinerant tribe of digital humanists caught in the middle of the rift. The latter shuttles back and forth between the two folds, mixing macro and micro scales of analysis, meshing the quantitative with the qualitative in the hope of creating sparks, frictions, a grand synthesis, a grand breakdown. With unimpeded access to ever-vaster cultural data sets, the separation may well grow or the process may produce generative synthesis. With an increasing number of platforms that combine qualitative analyses and quantitative methods, other outcomes are possible.

AN ALTERNATE CRISIS TAKES PLACE AS THE DIGITAL HUMANITIES BECOMES "THE HUMANITIES" *TOUT COURT*.

A CULTURE OF CRAFT RISES UP IN A POST-DIGITAL REVOLT THAT PRIVILEGES PHYSICAL PRESENCE OVER VIRTUAL PRESENCE, TOUCH OVER SIGHT AND SOUND, POOR MEDIA OVER RICH MEDIA.

Physical making, including self-consciously "backward" forms of manual work and handcrafting, has always accompanied and sometimes intersected digital culture. Such phenomena as the revival of knitting subcultures, the rise of a cottage industry of chapbook publishers, steampunk fabrication, makers fairs, even the Slow Food movement, were once confined to cities such as San Francisco and London, but are now spreading virally.

In the era of ubiquitous networks, they become the hubs of a revolt against screen culture, yet no distinction is made between the commercial Internet and the Digital Humanities. A significant sector of the humanities community splinters off and embraces these values as a critique of contemporary society, leaving their peers to bridge the two contexts, striving to conjugate the manual with the virtual, the macro with the micro, scholarship with arts-and-craft practice.

Regardless of our considerable enthusiasm for the social aspects and technological affordances of the Digital Humanities, there will always, and must always, be space for uninterrupted reading and reflection—such habits of mind stand in opposition to a culture that appears to demand multi-tasking and faceted attention at all times. The "classical" humanist attachment to concentration on the singular object, text, or task-at-hand may well become the mental equivalent of the attention to art and craft we now see in the realm of making.

AS CONCEPTS OF AUTHORSHIP, DOCUMENT, ARGUMENT, PROVENANCE, AND REFERENCE BECOME INCREASINGLY UNSTABLE, CONCEPTS THAT ARE FLUID, ITERATIVE, AND DISTRIBUTIVE, BUT LESS "AUTHORITATIVE," ARE TAKING THEIR PLACE.

YET IT IS BECOMING EASIER THAN EVER TO VALIDATE, TRACK, AND CROSS-CHECK INFORMATION.

Concepts of authorship in Digital Humanities research are already trending toward fluid, iterative, and distributive models. Whatever the medium, authorship is increasingly understood as a collaborative process, with individuals creating materials within the setting of a team that merges their identities into a corporate subject (the laboratory, the technology sandbox, the research group). Far from disappearing, authorial traces proliferate within the merged identity and can be brought to the surface by means of analytical tools that make it possible to track nano-units of authorship like isotopes of intellectual property whose fingerprints can be extracted from the swarm of discourse.

These fingerprints themselves will prove mutable as humanists become accustomed to working with flexible and modular discourse units and even embrace combinatoric writing styles. Generative processes of composition and algorithmic criticism could soon become widespread practices with the result that the old hypertext model of modules, units, nodes, and connections will return, but stripped of the need for elaborate menu-driven navigation systems. Rather than a garden of forking paths, readers and writers will enter combinatoric matrices where the associational trail and argument structures can be produced from Semantic Web capabilities in collaborations that are user-structured.

The notion of the document will shift accordingly. As the cumulative product of multiple interventions by multiple authors, the document migrates from medium to medium and platform to platform in ways that reshape its boundaries as a discursive object. A modular argument can be repurposed in chunks as small units of commentary weave a tight rhetorical web across a field of related artifacts, topics, or events. The field of reference thus becomes an emergent feature of discourse, one that is produced as an effect of interrelated arguments and exchanges as well as through the paratextual apparatus that spins outward with centrifugal force into the infinite inventory of precedents. A fully realized heteroglossic text will be a feature of technological and humanistic intervention, exhibiting the avenues and byways of associative trails in which the history and the future of composition are interwoven.

Specialized authority may be conceived less in terms of credential platforms such as universities than in terms of public performance, so that the scholarly expert and humanistic guide take their place alongside the imaginative storyteller as a conveyor of history and culture. Collaborative models of authorship, swarm writing, and collective production will make possible real-time integration of partial contributions into synthetic wholes. Expressions of groupthink, however, may obliterate some of the very foundations of autonomous thought, as writers embed themselves in social networks as part of the compositional process. Predictive models of creative and imaginative life may produce agonistic engagements that are generative and iterative in unexpected ways.

As every act of engagement with a digital world generates its own trail of data and metadata, the crucial tasks of forgetting, of strategically looking away, of ignoring, of letting go and even of erasure will become more critical. The practice of discrimination that distinguishes provenance and other features of reliability will need to be attended to by schoolchildren and scholars alike. The still-unanticipated effects of social media and their capacity for herd behavior and swarm politics in the realm of culture may shrink or may expand to the point of overwhelming any individual voice or talent except as a note of common reference in the shared field.

VISIONARY PARTNERSHIPS AMONG GOVERNMENTS, UNIVERSITIES, LIBRARIES, ARCHIVES, MUSEUMS, AND CULTURAL INSTITUTIONS HAVE THE POTENTIAL TO GIVE RISE TO A VAST DIGITAL CULTURAL COMMONS THAT SUPPORTS HUMANITIES RESEARCH IN THE PUBLIC INTEREST.

CITIZEN INVOLVEMENT IN THE CURATION, PRESERVATION, AND INTERPRETATION OF THE CULTURAL PATRIMONY WILL EXPAND.

BUT THIS CULTURAL COMMONS WILL CONTINUE TO BE RESTRICTED TO MATERIALS THAT ANTEDATE THE PAST 75 YEARS. THE PATRIMONY OF THE PRESENT ERA WILL BECOME A BATTLEGROUND BETWEEN THE ADVOCATES OF RESTRICTIVE VS. OPEN ACCESS, WITH DIGITAL HUMANISTS AT THE FOREFRONT OF THE OPEN-ACCESS MOVEMENT.

ADVANCED WORK IN THE DIGITAL HUMANITIES WILL EITHER TARGET PRE-CONTEMPORARY CULTURAL CORPORA OR LIVE DANGEROUSLY WITH RESPECT TO THE LAW.

Though frequent topics of discussion and debate, the great public works projects of the digital era have yet to be fully built: today's equivalents of the great road-building, electrification, and infrastructure development efforts of the industrial era are still unrealized. Where is the investment in the online equivalents of the Carnegie libraries, settlement houses, and other great philanthropic undertakings that promote the enfranchisement of all sectors of society? When and how will educational priorities change? Will our universities and colleges institutionalize approaches to learning and research grounded in collaboration and cooperation instead of celebrity and competition? Or will we continue to allow profit-driven entities to shape the networked environment on which our digital future depends? Will a cultural commons be established and made available to the citizens of the world from the privacy of their laptops? Or will distinct cultures emerge with their own rules and practices for use and access?

The creation of a global cultural commons has the potential to enhance the quality, depth, and reach of humanistic research. It also offers the prospect of resituating the humanities at the crossroads of contemporary public life. And it may well cast digital humanists in innovative public roles, create new audiences for cultural scholarship, and build bridges between the work of professional and amateur historians. Archival projects that make use of crowd-sourcing have begun to attract the participation of enthusiastic citizen scholars. Schoolchildren and their teachers are now able to work with primary-source materials and to make discoveries akin to those made by amateur astronomers studying the night sky. Communities create and curate their own archival resources, promoting cultural awareness and a sense of citizen ownership of the cultural patrimony.

The realization of such an inclusive vision faces a number of obstacles. None is more daunting than restrictions on the free circulation of the cultural patrimony of the past three-quarters of a century. Copyright restrictions have already led digital humanists to either focus their experiments in fields such as text-mining on 19TH century cultural materials or to boldly assert their right to the fair use of more recent materials, knowing full well that the university counsel's office would be unlikely to support their stance and that take-down orders may surface sooner or later. Because of copyright, such efforts must typically exclude the bulk of recent critical and scientific scholarship, even when such materials are accessible via online repositories. Even the non-consumptive use of digital assets remains a controversial matter and the object of negotiations with the owners of digital repositories. Entire fields of inquiry (on contemporary art, on certain authors, on huge swaths of popular culture, on topics where the primary assets are in corporate hands) remain off-limits or must operate either under the radar or be subject to exorbitant, arbitrary fees. Even the best-intentioned scholars are left to blindly navigate the murky waters of orphaned and protected works with conflicting understandings of ownership, permissions, and rights.

The copyright system is badly broken, and it is seriously curbing innovation on a multitude of fronts. Digital humanists will have no choice but to continue to storm the barricades for the causes of open access, copyright reform, and the global cultural commons. The future of the humanities and the "commons of the mind" depend upon the successful creation of such public spaces of knowledge production and knowledge exchange.

WORK IN THE HUMANITIES WILL RELY ON NEW MODES OF ASSESSMENT, ALTERED MODELS OF TRAINING, AND SHIFTS IN OUR UNDERSTANDING OF HOW WE VALUE PROFESSIONAL, CITIZEN, AND AMATEUR CONTRIBUTIONS TO KNOWLEDGE.

THE RISE OF CITIZEN SCHOLARS AND BIG HUMANITIES PROJECTS WILL AT ONCE BUILD BRIDGES BETWEEN THE ACADEMY AND SOCIETY AT LARGE AND REINFORCE FRICTIONS OVER THE SOCIAL ADVOCACY ROLES BEING PERFORMED BY PUBLIC HUMANITIES PROJECTS.

Whereas universities long ago developed standard practices for evaluating print-based humanistic scholarship, classroom teaching, and administrative service, Digital Humanities deals academic leaders a new hand of cards. The fact is that most digital projects are team-based; many are grant-driven (as in engineering and the sciences); they may involve partnerships with numerous extramural entities; they blur the boundary lines among research, teaching, and service; and they are iterative and may require extended timelines. Each of these fundamental aspects of digital projects creates additional social and disciplinary complexities. To these one must add the need to assess the intellectual value of outputs that may only partly correspond to traditional forms or genres of argument. Who should assess these? Who are the peer groups and what constitutes the community of expert evaluators? How ought design or technological inventiveness factor into professional reviews? Who credentials digital humanists? Is this a profession or merely an accessory set of capabilities?

Such properly "professional" questions are accompanied by others that relate to the involvement of non-university partners and to social outreach. Large-scale partnerships necessarily imply diminished control on the part of project directors as well as pressures to please "others" in order to secure continued cooperation. If the project director adopts a rigorously critical line while collaborating with a local historical society motivated by boosterism, that will surely result in reduced support; if, on the contrary, he or she adopts a flexible line, the project partner will be content, but the project may be dismissed by professional historians as "mere outreach." To what degree ought impact or visibility be considered a measure of success?

Many challenging issues lie ahead as regards the institutional life of Digital Humanities. Aside from the struggle for resources, there is an urgent need for a critical language to describe digital projects and for common—yet flexible—standards for evaluating animation, navigation, information architecture, and other features of born-digital projects and platforms.

STRATIFIED APPROACHES TO THE CONSERVATION OF HISTORICAL MATERIALS WILL DISPLACE THE UNIFORM CONSERVATIONIST IDEOLOGIES AND METHODOLOGIES OF THE 19TH CENTURY.

ERASURE AND FORGETTING WILL BECOME AS IMPORTANT TO THE HORIZONS OF HUMANISTIC WORK AS PRESERVATION AND REMEMBERING.

The centralized practices of collecting, processing, and preservation that developed over the past few centuries are responsible for the greatest act of historical recovery and retrieval in the history of humankind. Yet they also have led to an impasse. The sheer breadth and depth of materials now being collected, the growing volume of potential objects for collection and preservation, and the high degrees of redundancy characteristic of contemporary archival corpora have created a supply that so vastly exceeds the capabilities of institutions of memory that the result has been ever-burgeoning backlogs.

The solution is twofold. Nimble models of preservation, conservation, and processing must be developed that bring together researchers, archivists, curators, librarians, and members of the general public. Such models must reset all defaults with "quick and dirty" automated processing as the new norm, with full processing and preservation reserved for selected collections. Techniques such as automated metadata generation, user-tagging, and crowd-sourcing must be employed to expedite availability for user communities. The *user*-centered—not *document*- or *object*-centered—archive must become the rule. Gone is the era of the archive as a Fort Knox.

Total conservation and preservation is not an option, but it really never was. From the very start, institutions of memory were actively engaged in acts of selection and filtering that mostly took place behind closed doors. In the digital era, that process is being democratized, and digital humanists are being called upon to play a central role in archiving and curating collections, making decisions about preservation strategies, and critically reflecting upon the role of cultural forgetting and loss.

Digital archives lead a uniquely fragile existence, allayed only by redundancy and backups. Bit rot sets in; every act of transmission is a transformation; and no file is ever entirely self-identical. The recognition that the task of cultural memory is not exhaustive, but selective, that the shape of who we are is determined as much by what does not remain as what does, is a founding principle of humanistic scholarship and one that underscores the situated character of all knowledge. These principles will be challenged to confront the core tasks of collections-building.

Erasure studies will play as central a role in the future of the Digital Humanities as will collections-building, curation, interpretation, and annotation of and research on historical corpora.

A NEW KIND OF DIGITAL HUMANIST IS EMERGING WHO COMBINES IN-DEPTH TRAINING IN A SINGLE HUMANISTIC SUBFIELD WITH A MIX OF SKILLS DRAWN FROM DESIGN, COMPUTER SCIENCE, MEDIA WORK, CURATORIAL TRAINING, AND LIBRARY SCIENCE.

THE ZIGZAG DISPLACES LINEAR MODELS OF GRADUATE AND POST-GRADUATE TRAINING.

COMMUNITIES OF SCHOLARS ARE REORGANIZED ACCORDING TO DOMAINS OF PRACTICE, NOT ALONG DISCIPLINARY LINES.

THE COLLAPSE OF COMPREHENSIVE MODELS OF KNOWLEDGE (EVEN WITHIN SINGLE DISCIPLINES) BECOMES DEFINITIVE.

For the past half-century, comprehensive models of graduate training in the humanities, not unlike the sciences, have come under increasing pressure due to the explosion of subfields and specializations on the one hand, and the rise of new interdisciplinary fields on the other. The result is that the fiction of comprehensive training in even a single discipline survives on paper alone. In practice, doctoral students now establish themselves as experts in a specialized domain, which they then tie in to other intra- or interdisciplinary micro-domains, with so-called "theory" often serving as a bridge to broader conversations within the humanities or society at large.

The growing prominence of the Digital Humanities is introducing an additional set of pressures and complications as well as opportunities. The relatively linear tracks still being pursued in today's doctoral programs are already being displaced by zigzagging paths between applied and pure research; realms of doing, making and thinking; experiences of work as research and of research as work. "Outside" skills—skills in fields such as design, computer science, media practice, curation, or library science—are assuming increasing importance alongside core training in a given humanities specialty, with combinations established in a pragmatic, albeit ad hoc, manner. The result is greater variability in the professional profiles of young humanists, along with greater flexibility with respect to the job market. No longer trained for academic careers alone, skilled in practical as well as theoretical domains, they are moving more fluidly between institutions of memory, industry, and the academy. The Digital Humanities reframes our notion of the scholar from the tenured sage in a warren-like office to include a wider range of participants—staff members with research training, community archivists, curators of objects, designers who make it part of their practice to work on humanistic projects, programmers who specialize in cross-disciplinary tool building. This expansion of whom we think of as performing scholarship coincides, sometimes all too complicitly, with the de-tenuring of faculty and the inclusion of adjunct and precarious workers into every facet of academia. Here as elsewhere, the Digital Humanities can be used to justify either the best or the worst of intentions, making it incumbent upon those who would adopt the mantle of digital humanist to do so mindful of the pitfalls as well as the promises.

The prevailing research culture of the Digital Humanities will become entrepreneurial, much like design or certain areas of contemporary engineering and the sciences are. Careers will be built around answers to questions like: Where do the most interesting opportunities lie? What are the richest archival repositories? What is fundable? Which lab is doing the most exciting work? The digital humanist's sense of identity will be less anchored in a discipline or disciplinary specialty than in a sense of belonging to a community of practice within which tools and methods are primary and objects of study are secondary considerations.

DESIGN EMERGES AS THE NEW FOUNDATION FOR THE CONCEPTUALIZATION AND PRODUCTION OF KNOWLEDGE.

DESIGN METHODS INFORM ALL ASPECTS OF HUMANISTIC PRACTICE, JUST AS RHETORIC ONCE SERVED AS BOTH ITS GLUE AND COMPOSITIONAL TECHNIQUE.

CONTEMPORARY ELOQUENCE, POWER, AND PERSUASION MERGE TRADITIONAL VERBAL AND ARGUMENTATIVE SKILLS WITH THE PRACTICE OF MULTIMEDIA LITERACY SHAPED BY AN UNDERSTANDING OF THE PRINCIPLES OF DESIGN.

As the well-oiled machinery of print culture finds itself jammed by the volatile intermedia mix of the digital era, the form that knowledge assumes can no longer be considered a given. Knowledge-making and knowledge design become radically intertwined endeavors: so much so that digital humanists increasingly find themselves called upon to operate as and/or collaborate with designers.

Design means shaping knowledge and endowing it with form; the field of design encompasses structures of argument. We have discussed how the capacious umbrella of "design" incorporates a wide variety of practices: project design, the design of database architectures, metadata schemes, graphic and typographic design, user interface design, data visualization, information architectures, interactivity design, and the crafting of narrative and argumentative structures in multiple media. No digital humanist can become proficient in all, but every digital humanist will have to become familiar with all. The reason is simple: As we hope we have already demonstrated, digital projects of any scope require teams, not individuals, for each phase of design development and implementation. Developing an understanding of the ways the technical components of a project mesh is just as essential as mastering specific skills.

The central role played by design implies new challenges as well as new opportunities. New challenges because design dexterity requires specialized skills that lie outside the traditional knowledge-base of humanists. Amateurish design plagued many early Digital Humanities experiments and contributed to their premature demise.

All future scholarly projects that do not aspire to the highest design standards are unlikely to achieve public impact or enduring results. This is the reason why meticulous attention to design also provides new opportunities: namely, that good design breeds rich and robust digital tools and resources, and can make specialized forms of knowledge and inquiry comprehensible to expanded audiences and user groups.

Digital humanists still have much to learn about the design and production of networked repositories, systems of communication, and new media environments. Modeling knowledge using digital tools and platforms provides a powerful perspective from which to engage in critical analysis of the rhetorical force and ideological shape of these very modes. Practice and theory inform each other in the process of making. Without making, theory has no traction. Without theory, practice has no critical purchase.

Early practitioners of Digital Humanities were willing to tinker with technology and conquer a steep learning curve in acquiring technical skills. Though much groundwork has been laid in the field of digitally based scholarship, for innovation to occur, humanists have to be *inside* the technology, ready to plunge into the workings of platforms and protocols at least enough to understand how to think critically and imaginatively regarding the tools they employ. Technical tools and research questions are not unrelated. Coding can be as mindless as any other task, and knowing how to make things work does not guarantee insight. But ideological critique and critical studies of media also reach a limit without some knowledge of technological underpinnings.

The time of diagrammatic thinking is upon us. We need graphical interfaces for multidimensional and multimedia authoring that take advantage of computers' abilities to aggregate, synthesize, and organize arguments along multiple axes. Authorship and display must converge in such a way that arguments become visible and can be made both graphically and spatially. Relations among visible entities, as well as verbal units of thought, become tractable in the process. Ways of describing relations and visually structuring arguments through juxtaposition, derivation, hierarchy, equivalence, and other spatial relational concepts will introduce an interpretive dimension and enrich understandings of information design in the process.

The glass wall of the screen must become malleable. Crafting arguments in digital form and out of objects belonging to the full spectrum of media types must become no less fluid than doodling with a felt-tip pen on a paper sketch pad.

IF THE HUMANITIES ARE TO THRIVE AND NOT JUST EXIST IN NICHES OF PRIVILEGE, THEY WILL HAVE TO VISIBLY DEMONSTRATE THE CONTRIBUTIONS TO KNOWLEDGE AND SOCIETY THEY ARE MAKING IN THE DIGITAL ERA.

THIS MEANS SHAPING—NOT PARROTING OR SIMPLY USING—THE LANGUAGE OF OUR ERA.

However contradictory and heterogeneous, plausible or implausible, the scenarios sketched out in this concluding section raise fundamental questions about the sort of vision that could or should shape Digital Humanities as it builds the present and looks toward the future. The vision will have to be bold if the humanities are not to recede into an ornamental role within contemporary research universities. More than just being conversant with the defining languages of our epoch, the humanities will have to prove capable of informing those very languages.

The conviction that animates this book is that Digital Humanities is well-equipped to take on this task as it enters the mature phase of its existence. Understood as a critical experimental practice, carried out in the public laboratory of a cultural commons, Digital Humanities is itself a work-in-progress as much as a future promise, driving digital tool- and platform-development with content-specific research questions that design, investigate, and interrogate the cultural record of humanity.

Will Digital Humanities save the humanities in an era when traditional humanistic forms of inquiry and discourse find themselves drowned out by the din of commerce, the drumbeat of the 24-hour news cycle, and rampant tides of economism and vocationalism? Time will tell. But it is worth recalling that consumerism and perpetual information overload are but the flip side of an era in which both experts and ordinary citizens have unprecedented access to information. Literacy has taken on varied forms, and culture industries are flourishing on a scale that would have been unimaginable only a century ago. These expansions challenge conventions of cultural value and canon formation. This much is knowable: The future course of the humanities will hinge upon informed and imaginative engagement with the historical forces that are shaping our times, our communities, and ourselves.

This final section of *Digital_Humanities* reflects on the preceding chapters, but can also stand alone as a concise overview of the field. As digital methodologies, tools, and skills become increasingly central to work in the humanities, questions regarding fundamentals, project outcomes, assessment, and design have become urgent. The specifications provide a set of checklists to guide those who do work in the Digital Humanities, as well as those who are asked to assess and fund Digital Humanities scholars, projects, and initiatives.

A SHORT GUIDE TO THE DIGITAL_HUMANITIES

QUESTIONS & ANSWERS

SPECIFICATIONS

QUESTIONS & ANSWERS I DIGITAL HUMANITIES FUNDAMENTALS

What is the Digital Humanities?

Digital Humanities refers to new modes of scholarship and institutional units for collaborative, transdisciplinary, and computationally engaged research, teaching, and publication.

Digital Humanities is less a unified field than an array of convergent practices that explore a universe in which print is no longer the primary medium in which knowledge is produced and disseminated.

Digital tools, techniques, and media have expanded traditional concepts of knowledge in the arts, humanities and social sciences, but Digital Humanities is not solely "about" the digital (in the sense of limiting its scope to the study of digital culture). Nor is Digital Humanities only "about" the humanities as traditionally understood since it argues for a remapping of traditional practices. Rather, Digital Humanities is defined by the opportunities and challenges that arise from the conjunction of the term *digital* with the term *humanities* to form a new collective singular.

The opportunities include redrawing the boundary lines among the humanities, the social sciences, the arts, and the natural sciences; expanding the audience and social impact of scholarship in the humanities; developing new forms of inquiry and knowledge production and reinvigorating ones that have fallen by the wayside; training future generations of humanists through hands-on, project-based learning as a complement to classroom-based learning; and developing practices that expand the scope, enhance the quality, and increase the visibility of humanistic research.

The challenges include addressing fundamental questions such as: How can skills traditionally used in the humanities be reshaped in multimedia terms? How and by whom will the contours of cultural and historical memory be defined in the digital era? How might practices such as digital storytelling coincide with or diverge from oral or print-based storytelling? What is the place of *humanitas* in a networked world?

What defines the Digital Humanities now?

The computational era has been underway since World War II, but after the advent of personal computing, the World Wide Web, mobile communication, and social media, the digital revolution entered a new phase, giving rise to a vastly expanded, globalized public sphere and to transformed possibilities for knowledge creation and dissemination.

Building on the first generation of computational humanities work, more recent Digital Humanities activity seeks to revitalize liberal arts traditions in the electronically inflected language of the 21st century: a language in which, uprooted from its long-standing paper support, text is increasingly wedded to still and moving images as well as to sound, and supports have become increasingly mobile, open, and extensible.

And the notion of the primacy of text itself is being challenged. Whereas the initial waves of computational humanities concentrated on everything from word frequency studies and textual analysis (classification systems, mark-up, encoding) to hypertext editing and textual database construction, contemporary Digital Humanities marks a move beyond a privileging of the textual, emphasizing graphical methods of knowledge production and organization, design as an integral component of research, transmedia crisscrossings, and an expanded concept of the sensorium of humanistic knowledge. It is also characterized by an intensified focus on the building of transferrable tools, environments, and platforms for collaborative scholarly work and by an emphasis upon curation as a defining feature of scholarly practice.

What isn't the Digital Humanities?

The mere use of digital tools for the purpose of humanistic research and communication does not qualify as Digital Humanities. Nor, as already noted, is Digital Humanities to be understood as the study of digital artifacts, new media, or contemporary culture in place of physical artifacts, old media, or historical culture.

On the contrary, Digital Humanities understands its object of study as the entire human record, from prehistory to the present. This is why fields such as classics and archaeology have played just as important a role in the development of Digital Humanities as has, for example, media studies. This is also why some of the major sectors of Digital Humanities research extend outside the traditional core of the humanities to embrace quantitative methods from the social and natural sciences as well as techniques and modes of thinking from the arts.

Where does the Digital Humanities come from?

The roots of computational work in the humanities stretch back to 1949 when the Jesuit scholar Roberto Busa, working in collaboration with IBM, undertook the creation of an automated approach to his vast *Index Thomisticus*, a computer-generated concordance to the writings of Thomas Aquinas. By means of such early uses of mainframe computers to automate tasks such as word-searching, sorting, counting, and listing, scholars could process textual corpora on a scale unthinkable with prior methods that relied on handwritten or typed index cards. Other early projects included the debut, in 1966, of *Computers and the Humanities*, the first specialized journal in the field. Seven years later, the Association for Literary and Linguistic Computing (ALLC) was founded, with the Association for Computers and the Humanities (ACH) following in 1978.

By the mid-1980s computational methods for linguistic analysis had become widespread enough that protocols for tagging digital texts were needed. This spurred the development of the Text Encoding Initiative (TEI). This important undertaking reshaped the field of electronic textual scholarship and led subsequent digital editing to be carried out in Extensible Markup Language (XML), the tag scheme of which TEI is a specialized subset. The first humanities-based experiments with database structures and hypertextual editing structured around links and nodes (rather than the linear conventions of print) date from this period, as do the many pilot projects in computational humanities in the United States sponsored by the National Endowment for the Humanities and other agencies, organizations, and foundations.

How do the Web and other networks affect the Digital Humanities?

As this revolution in protocols was taking place, the explosion of personal computing in the mid-1980s combined with the advent of the World Wide Web a decade later gave rise to a new generation of Digital Humanities work that was less text-centered and more design-driven. The desktop environment—with its graphical user interface, real-time WYSIWYG toolkit, and evolution from command lines to icons and window-based frames—not only vastly expanded the corpus of born-digital documents but also ushered in the gradual integration of audio, video, and graphics.

This integration has matured over the past decades and given Web culture its profoundly multimedial character. It also favored the enhancement of models of sharing, co-creation, publication, and community-building that have situated the Web at the center of contemporary social debates and socio-economic processes. The concept of the Web as a public sphere that extends the physical public spaces of contemporary life has, of course, been intensified thanks to smartphones, tablets, and other ubiquitous and pervasive computing and media devices.

What is ahead for the Digital Humanities?

Contemporary Digital Humanities stands not in opposition to the past, but on its shoulders. It honors the pioneering labors carried out over the past seven decades in the form of statistical processing (computational linguistics), linking (hypertext), modeling (architectural and visual displays), the creation of structured data (XML), and iterative editing and version control (for critical editions as well as analysis and creative practices), even as it seeks to move beyond repository building and editing to new synthetic practices. It is inspired by the same core conviction that animated computational humanities and early Digital Humanities pioneers: the conviction that computational tools have the potential to transform the content, scope, methodologies, and audience of humanistic inquiry.

QUESTIONS & ANSWERS 2
THE PROJECT AS BASIC UNIT

Why projects?

Projects are both nouns and verbs: A project is a kind of scholarship that requires design, management, negotiation, and collaboration. It is also scholarship that projects, in the sense of futurity, as something which is not yet. Projects are often pursued in teams, with collaborators bringing complementary skill-sets and interests to conceptualize the research questions being investigated and design possible trajectories for them to be answered. Hence, projects are projective, involving iterative processes and many dimensions of coordination, experimentation, and production.

Who is involved in Digital Humanities projects?

Digital Humanities projects typically involve multiple circles of researchers, from faculty and staff to students and community partners. A project's complexity and scale generally implies the involvement of multiple strata of personnel from within and across institutions of learning.

Projects can involve partner institutions such as museums, libraries, and archives as well as members of the community, alumni, and members of interested virtual networks such as collectors, amateur historians, and the like.

Partnerships with corporations, in particular media and technology companies, are also possible, with a caveat that corporate and academic cultures may be different in their goals and values.

How are Digital Humanities projects organized?

Projects are usually faculty-, staff-, or student-initiated. They are often built around a research question and/or a university collection or archival repository. Many take place outside the classroom; others involve a research project that is anchored in a recurring course.

A Principal Investigator (or, PI), co-PIs, project advisors, staff, interns, and students are all part of the project team. It is the responsibility of the PI to organize the project team, establish timelines for deliverables, and assess the project at each stage of development.

What is the difference between Digital Humanities projects and Big Humanities projects?

Digital Humanities projects come in all sizes: big, medium, and small. Some of the defining early Digital Humanities projects, however, as well as prominent contemporary work have assumed the form of Big Humanities projects, which are realized over many years, with many contributors, developers, and funders involved at various stages of development. Big Humanities projects are built along the lines of Big Science. They involve large-scale, long-term, team-based initiatives that build big pictures out of the tesserae of expert knowledge. The researchers and team members, from historians to technologists to designers, may number in the hundreds.

Little or "lowercase" Digital Humanities projects are typically carried out by individuals or small teams in consultation with experienced staff. As standard platforms and protocols have emerged, editing, exhibit-building, network analysis, and repository development require less one-off investment.

The bulk of Digital Humanities projects fall in between the two ends of the spectrum.

How is the Digital Humanities continuous with traditional forms of research and teaching in the humanities?

Like traditional humanities-based research and teaching, Digital Humanities work involves practices of analysis, critique, and interpretation; editing and annotation; historical research and contextualization. It examines the formal and historical properties of works of the imagination, the interplay of self and society, the history of ideas and of material culture. It attends to qualitative and non-quantifiable features of the human experience: complexity, ambiguity, medium specificity, and subjectivity. It builds on traditional approaches to the study, preservation, and classification of cultural corpora.

Though the range of media with which Digital Humanities works extends beyond the textual, its core commitments harmonize with the long-standing values of the humanistic tradition: the pursuit of analytical acuity and clarity, the making of effective arguments, the rigorous use of evidence, and communicative expressivity and efficacy. Digital Humanities then melds hands-on work with vastly expanded data sets, across media and through new couplings of the digital and the physical, resulting in definitions of and engagements with knowledge that encompass the entire human sensorium.

Both the traditional classroom and solitary study remain key features in the landscape of Digital

Humanities learning. At the same time, many precedents for collaborative work in communities of letters and knowledge networks are enhanced by digital platforms in a fabric animated by opportunities for hands-on, project-based learning. Since antiquity, the dominant models of humanistic inquiry have favored an understanding of intellectual labor as solitary and contemplative, cut off from—and even superior to—manual labor and the realm of making or doing. Digital Humanities re-embeds these models in an augmented model of pedagogy that emphasizes learning through making and doing, whether on the level of the individual or the group.

How is the Digital Humanities discontinuous with traditional forms of research and teaching in the humanities?

For nearly six centuries, humanistic models of knowledge have been shaped by the power of print as the primary medium of knowledge production and dissemination. Rather than rejecting print culture or embracing the simple pouring of print models into digital molds, Digital Humanities is engaged in developing print-plus and post-print models of knowledge. Both involve more than an updating of the knowledge delivery system. They entail the cognitive and epistemological reshaping of humanistic fields as a function of the affordances provided by the digital with respect to print. They also respect the increasing role teamwork and collaboration play in humanities research and training.

How does the Digital Humanities function in the print-plus era?

Print typically offers a single viewing angle, linear organization, a research output characterized by finitude and stability, and a scale of documentation and argumentation that has to respect the physical proportions of the book. The digital print-plus era, in contrast, allows for toggling back and forth between multiple views of the same materials. It allows for fluid scale shifts, for "zooming" from the macro- to the micro-level, and for the interweaving of data sets (such as source materials, notes, and correspondence) into research outputs. The screens and augmented spaces of the print-plus era allow for the faceting, filtering, and versioning of corpora; for the coexistence of multiple pathways within a single repository; for multilinear forms of argument. It is extensible in the double sense of allowing for seemingly unlimited scale and of being process- rather than product-based. When a book goes to print, it stabilizes in an edition that has to be reissued in order to be revised; a digital artifact can be altered or revised on a rewritable substrate that supports rapid refresh rates. The same digital artifact can lead multiple lives on multiple platforms, with multiple authors. It can undergo remixing by others before, during, and after its "completion."

How are Digital Humanities projects funded and sustained?

Because they cross over boundaries between disciplines; between theoretical and applied knowledge; and among the humanities, library science, information technology, and design, Digital Humanities projects typically require support structures that cut across conventional department and school organizational lines. Private foundations, public granting agencies, and industry partners have all provided monies for projects at every scale.

Funding for research in the humanities is far more limited than in the science and engineering fields, but the scope and innovative character of the Digital Humanities have led many projects to successfully garner external funding. In order to attract and sustain such funding, it has proven essential for projects to receive internal support during a period of incubation so that they may prove their worth by successfully reaching an initial set of benchmarks.

Sustaining such projects requires that faculty and students who assume leadership positions need the support and recognition that this work is a combination of research, teaching, and service.

What are the prevailing crediting and attribution conventions and authorship models for Digital Humanities projects?

Traditional authorship and crediting practices in the humanities are based on single authorship. Although practices of attribution are still fluid in the Digital Humanities community, the emerging model recognizes that many, if not most, Digital Humanities projects are analogous either to natural science laboratory projects or to the collaborative attribution system used in the performing arts.

No standardized crediting system for Digital Humanities projects has been embraced universally. But the dominant trend is toward the differentiation of roles such as principal investigator, researcher, designer, programmer, modeler, editor, and the like.

INSTITUTIONS AND PRAGMATICS

How do Digital Humanities projects interconnect the classroom with libraries, museums, and archives?

Most colleges and universities have extensive resources for research and study that are underutilized after fulfilling their core research, teaching, and training missions. Contemporary Digital Humanities taps these riches by expanding the concept of the classroom to encompass library, museum, and archival collections, positioning them as central training places via hands-on research in the company of peers.

Much as in a natural science laboratory, students involved in Digital Humanities projects learn by making and doing, working within this extended classroom under the guidance of expert curators, archivists, and researchers, and in the company of peers. Whereas traditional models of humanistic training view the acquisition of skill-sets and disciplinary training as preconditions for the transition to becoming engaged in the creation of original scholarship, Digital Humanities work accelerates this apprenticeship, inserting students into research communities from the start.

How can Digital Humanities projects involve inter-university collaboration?

The scale and scope of many Digital Humanities projects, as well as their ties to physical collections and IT infrastructure needs, make them ideally suited to inter-university collaboration. Projects can be developed and divided up strategically among multiple partner institutions leveraging specific strengths, distributing workloads, sharing the benefits of research outcomes, and building cross-institutional bridges.

Benefits include cost-sharing and enhanced prospects for external funding. But they also transcend the practical sphere: They enable Big Humanities models of research whose outcomes are of potential interest to broad cross-disciplinary and nonspecialist audiences. By involving multiple institutions, such projects contribute to a sense of shared identity and of belonging to a broader research community. They also help to answer endemic student anxieties regarding the practical value of humanities knowledge and research.

How can Digital Humanities projects involve expertise outside the academy?

Many Digital Humanities projects develop entirely within a single college or university. But others require domains of knowledge and forms of expertise that are under- or unrepresented in or lie outside the confines of academic fields. Combining intra- and extramural expertise within well-designed Digital Humanities projects often proves essential to their success. Such approaches include work with communities of collectors and historical associations and the use of crowd-sourcing for the processing, transcription, and annotation of archival documents. Not only can the scope and quality of humanities research benefit from such partnerships, but they also contribute to the creation of a new class of citizen scholars who otherwise would be mere citizen consumers.

How can extramural partnerships play a role in developing, supporting, and sustaining Digital Humanities projects?

The promotion of public knowledge is a core value of the Digital Humanities. Extramural partnerships—whether with professional societies, historical associations, institutions of informal learning (libraries, museums, archives), corporations, or public entities—can extend the reach and impact of humanities research in contemporary society. The most successful partnerships address questions of shared critical interest with research results that rise to the highest standards of scholarly rigor while being conjugated across multiple media platforms in the "language" of the partner institutions through exhibitions, performances, books, Web publications, or other means.

Partnerships can expand the depth and diversity of the talent pool of available participants in a project, broaden a project's potential audience and impact, and, as with inter-university collaborations, help to solidify short- and long-term financial sustainability.

How can educational institutions support Digital Humanities research?

Digital Humanities research projects require fluid boundary lines among academic departments and institutional units. Because the projects are often team-based and imply merged models of theoretical and applied knowledge across the traditionally separated domains of "research," "teaching," and "service," elements such as design facilities, information systems, multimedia production, IT work, and collections-based research are not mere "supports," but rather integral features of project design and execution.

In addition to promoting a culture where such boundary lines do not stand in the way of innovation, institutions must embrace co-teaching as a standard feature of the new landscape of the humanities, rather than penalizing it as a form of work reduction. Co-creation must be seen as a legitimate form of scholarly and student intellectual labor, complementary to traditional forms of output. The easing of access and use-restrictions on museum, archive, and library special collections represents a key precondition to the creation of an expanded, hands-on classroom, and serves open-access models equating preservation with proliferation, rather than restricted control.

College and university legal offices must be careful not to interpret copyright restrictions narrowly out of an unwillingness to broker hypothetical risks. Fair use needs to be understood in the broadest possible sense in order not to shackle Digital Humanities research. College- and university-based collections need to be shared with the research community as freely as possible.

Last but not least, institutions of higher learning must promote and foster a less risk-averse culture in the humanities disciplines: a culture where, as in the sciences, "failure" would be accepted as a productive outcome when undertaking innovative, speculative work. Differentiating between productive forms of failure and poor research is essential to promoting research communities where innovation is a core value.

What are the institutional niches that best support Digital Humanities projects?

Digital Humanities projects have generally flourished less within single departments, schools or institutional units, than across such structures. Even humanities research centers, built to house and support the research of individual scholars, have not always proven to be the ideal home—although some have successfully reshaped their policies, funding models, and physical infrastructure to support collaborative Digital Humanities work.

More typically, Digital Humanities has thrived in independent, free-standing laboratories or centers where there exists a community of scholars (humanists and non-humanists alike), staff members, curators, and students interested in the shared exploration of innovative models of scholarship. Such environments are best envisaged as a hybrid of making, thinking, and play spaces, combining computational facilities; digital imaging, sound, and video production facilities; and meeting and exhibition spaces.

How can institutions assess the scale of investment and expectation for Digital Humanities projects appropriate to them?

Projects come in all sizes. There is no inherent reason why a large project cannot be undertaken by a small institution or a small project by a large institution. Nor is there any inherent reason why individual scholars cannot undertake large-scale collaborations among multiple colleges or universities.

So there is no single formula for success. The scale and form that Digital Humanities projects take must be dictated by thoughtful project design—combining research questions, ambitions, and anticipated outputs—as well as the available logistical, personnel, and financial resources. Much as in the laboratory sciences, this implies a balance between pragmatic vision and entrepreneurial initiative.

How can peers and academic leaders assess Digital Humanities projects?

Metrics for evaluating the quality and impact of Digital Humanities projects combine traditional assessment methods in the humanities with new factors. Peer review remains fundamental to processes of assessment, but now draws as much from the community of leading Digital Humanities practitioners as from field-based peers. A less risk-averse culture is the prerequisite for a more innovation- and experimentation-driven model of the Digital Humanities to take hold.

In addition to traditional peer-based criteria, some assessment tools that have a long history in the natural and social sciences may become relevant to humanities fields: citations, grant-writing success, public impact, and the like. It should be noted that variations in the sizes of fields make caution essential in the use of quantitative tools; otherwise they will provide very crude, and possibly misleading, measures of importance or impact. Original scholarship and intellectual rigor remain the essence of Digital Humanities work.

Traditional print-based metrics of productivity are already being eclipsed by the realities of print-plus and digital publishing, so expectations of productivity must encompass multiple media, different formats, and variable scales of contributions to knowledge. In other words, the media and technologies in which intellectual work is realized matter as much as its "content." This means that the "work" is not just the content but, rather, *everything*: the environment that has been designed for the work's performance and publication; the interface and data structures, the back-end database, and the code that enables multiple forms of audience engagement. All of these matter in assessments of quality and rigor.

SPECIFICATION I
HOW TO EVALUATE DIGITAL SCHOLARSHIP

This text provides a set of guidelines for the evaluation of digital scholarship in the humanities, social sciences, arts, and related disciplines. The guidelines are aimed, foremost, at academic review committees, chairs, deans, and provosts who want to know how to assess and evaluate digital scholarship in the hiring, tenure, and promotion process.
The list is also intended to inform the development of institution-wide policies for supporting and evaluating scholarship and creative work that reflects traditional values while incorporating specific understandings of new platforms and formats.

Fundamentals for initial review

The work must be evaluated in the medium in which it was produced and published. If it is a website, that means viewing it in a browser with the appropriate plug-ins necessary for the site to work. If it is a virtual simulation model, that may mean going to a laboratory outfitted with the necessary software and projection systems to view the model. Work that is time-based—such as videos—will often be represented by stills, but reviewers also need to devote attention to clips in order to fully evaluate the work. The same can be said for interface development, since still images cannot fully demonstrate the interactive nature of interface research. Authors of digital works should provide a list of system requirements (both hardware and software, including compatible browsers, versions, and plug-ins) for viewing the work. It is incumbent upon academic personnel offices to verify that the appropriate technologies are available and installed on the systems that will be used by the reviewers before they evaluate the digital work.

Crediting

Digital projects are often collaborative in nature, involving teams of scholars who work together in different venues over various periods of time. Authors of digital works should provide a clear articulation of the role or roles that they have played in the genesis, development, and execution of the digital project. It is impractical—if not impossible—to separate out every micro-contribution made by team members since digital projects are often synergistic, iterative, experimental, and even dynamically generated through ongoing collaborations. Nevertheless, authors should indicate the roles that they played (and time commitments) at each phase of the project development. Who conceptualized the project and designed the initial specifications (functional and technical)? Who created the mock-ups? Who wrote the grant proposals or secured the funding that supported the project? What role did each contributor play in the development and execution of the project? Who authored the content? Who decided how that content would be accessed, displayed, and stored? What is the "public face" of the project and who represents it and how?

Intellectual rigor

Digital projects vary tremendously and may not "look" like traditional academic scholarship; at the same time, scholarly rigor must be assessed by examining how the work contributes to and advances the state of knowledge in a given field or fields. What is the nature of the new knowledge created? What is the methodology used to create this knowledge? It is important for review committees to recognize that new knowledge is not just new content but also new ways of organizing, classifying, and interacting with content. This means that part of the intellectual contribution of a digital project is the design of the interface, the database, and the code, all of which govern the form of the content. Digital scholars are not only in the position of doing original research but also of inventing new scholarly platforms. Five hundred years of print have so fully naturalized the "look" of knowledge that it may be difficult for reviewers to fully understand these new forms of documentation and the intellectual effort that goes into developing them. This is the dual burden—and the dual opportunity—for creativity in the digital domain.

Crossing research, teaching, and service

Digital projects almost always have multiple applications and uses that enhance research, teaching, and service. Digital research projects can make transformative contributions in the classroom and sometimes even have an impact on the public-at-large. This ripple effect should not be diminished. Review committees need to be attentive to colleagues who dismiss the research contributions of digital work by cavalierly characterizing it as a mere "tool" for teaching or service. Tools shape knowledge, and knowledge shapes tools. But it is also important that review committees focus on the research contributions of the digital work by asking questions such as the following: How is the work engaged with a problem specific to a scholarly discipline or group of disciplines? How does the work reframe that problem or contribute to a new way of understanding the problem? How does the work advance an argument through both the content and the way the content is presented? How is the design of the platform an argument? To answer this last question, review committees might ask for documentation describing the development process and design of the platform or software, such as database schemata, interface designs, modules of code (and explanations of what they do), as well as sample data types. If the project is, in fact, primarily for teaching, how has it transformed the learning environment? What contributions has it made to learning and how have these contributions been assessed?

Peer review

Digital projects should be peer-reviewed by scholars in fields who are able to assess the project's contribution to knowledge and situate it within the relevant intellectual landscape. Peer review can happen formally through letters of solicitation but can also be assessed through online forums, citations, and discussions in scholarly venues, by grants received from foundations and other sources of funding, and through public presentations of the project at conferences and symposia. Has the project given rise to publications in peer-reviewed journals or won prizes by professional associations? How does it measure up to comparable projects in the field that use or develop similar technologies or similar kinds of data? Finally, grants received are often significant indicators of peer review. It is important that reviewers familiarize themselves with grant organizations across schools and disciplines, including the humanities, the social sciences, the arts, information studies and library sciences, and the natural sciences, since these are indicators of prestige and impact.

Impact

Digital projects can have an impact on numerous fields in the academy as well as across institutions and even the general public. They often cross the divide that arises among research, teaching, and service in innovative ways. Impact can be measured in many ways, including the following: support by granting agencies or foundations, number of viewers or contributors to a site and what they contribute, citations in both traditional literature and online (blogs, social media, links, and trackbacks), use or adoption of the project by other scholars and institutions, conferences and symposia featuring the project, and resonance in public and community outreach (such as museum exhibitions, public policy impact, adoption in curricula, and so forth).

Approximating equivalencies

Is a digital research project "equivalent" to a book published by a university press, an edited volume or a research article? These sorts of questions are often misguided since they are predicated on comparing fundamentally different knowledge artifacts and, perhaps more problematically, consider print publications as the norm and benchmark from which to measure all other work. Reviewers should be able to assess the significance of the digital work based on a number of factors: the quality and quantity of the research that contributed to the project; the length of time spent and the kind of intellectual investment of the creators and contributors; the range, depth, and forms of the content types and the ways in which this content is presented; and the nature of the authorship and publication process. Large-scale projects with major funding, multiple collaborators, and a wide-range of scholarly outputs may justifiably be given more weight in the review and promotion process than smaller-scale or short-term projects.

Development cycles, sustainability, and ethics

It is important that review committees recognize the iterative nature of digital projects, which may entail multiple reviews over several review cycles, as projects grow, change, and mature. Given that academic review cycles are generally several years apart (while digital advances occur more rapidly), reviewers should consider individual projects in their specific contexts. At what "stage" is the project in its current form? Is it considered "complete" by the creators, or will it continue in new iterations, perhaps through spin-off projects and further development? Has the project followed the best practices, as they have been established in the field, in terms of data collection and content production, the use of standards, and appropriate documentation? How will the project "live" and be accessible in the future, and what sort of infrastructure will be necessary to support it? Here, project specific needs and institutional obligations come together at the highest levels and should be discussed openly with deans and provosts, library and IT staff, and project leaders. Finally, digital projects may raise critical ethical issues about the nature and value of cultural preservation, public history, participatory culture and accessibility, digital diversity, and collection curation which should be thoughtfully considered by project leaders and review committees.

Experimentation and risk-taking

Digital projects in the humanities, social sciences, and arts share with experimental practices in the sciences a willingness to be open about iteration and negative results. As such, experimentation and trial-and-error are inherent parts of digital research and must be recognized. The processes of experimentation can be documented and can prove to be essential in the long-term development process of an idea or project. White papers, sets of best practices, new design environments, and publications can result from such projects, and these should be considered in the review process. Experimentation and risk-taking in scholarship represent the best of what the university, in all its many disciplines, has to offer society. To treat scholarship that takes on risk and the challenge of experimentation as an activity of secondary (or no) value for promotion and advancement can only serve to reduce innovation, reward mediocrity, and retard the development of research.

SPECIFICATION 2
PROJECT-BASED SCHOLARSHIP

Project-based scholarship exemplifies contemporary Digital Humanities principles. It differs from traditional scholarly publication in being team-based, distributed in its production and outcome, dependent on networked resources (technical and/or administrative), and in being iterative and ongoing, rather than fixed or final, in its outcome. It necessarily involves many dimensions of conception, design, coordination, and resource use that build extra layers of complexity onto the traditional approach to humanities research. The following list is useful to the creation of a grant proposal or research plan for project-based work and reflects best-practices standards (with the caveat that debate persists).

Contribution to knowledge

The project should meet the criteria of any scholarly work through its contribution to knowledge in a discipline or field. How is the project in dialogue with an issue or topic in a given disciplinary field and how does it move the discourse forward in an innovative way? Does the project contribute to and advance the state of knowledge of a given field or fields?

The model of knowledge

How is the knowledge shaped and modeled: as an argument, a presentation, a display? What can be taken from the project as a theoretical principle, method, or information that is useful for other scholars, including those who are not engaged with Digital Humanities research? How does the project model and embody new knowledge?

Research questions and digital media

Digital environments allow for different approaches for relating and processing materials and this should be demonstrated in the research plan. Simply putting something online is not digital research. The litmus test is to ask what is being done that could not be done in print-based or traditional scholarship. How has the research project been formulated from within the affordances of digital methods?

Tools and content

Many digital projects involve innovative recombining and reconfiguring of existing tools toward the formulation of new knowledge. Is this a tools-based project or a content-driven project and how do these intersect? How can the intellectual labor of the design and development of the "tool" be assessed in tandem with the "content"? To what extent are they inextricable and why?

Methods

Does the project have a thesis or guiding methodological principle? How did the digital platform allow it to be explored, tested, argued, demonstrated, or even refuted?

Born digital and/or digitized artifact

Digital projects often combine analog materials that have been scanned or digitized and elements that are born digital—analysis, research, processing, or newly authored files. Elements of information structure are also born digital. How are each of these elements understood and what role do they play in the overall project?

Collections-sharing and licensing

The future of humanistic learning and the level of societal impact that humanities scholarship can achieve depend upon unrestricted access to cultural and historical repositories; accordingly, the least restrictive licenses should be the norm. What kinds of licensing and intellectual property issues will the project encounter? How can the work be accessed and used by the scholarly community and public-at-large?

Interface as knowledge representation and content-modeling

The interface of a project expresses an argument in its design. Does it offer a snapshot of the contents of the project, or a set of entry points for activities that can be performed? Understanding the ways the interface is structured, how it embodies the ideas of the project, and how it supports the engagement with the project is essential.

Team, collaborative, and project management

Knowing who will take responsibility for each part of a digital project is crucial for development and design. Each participant's role should be spelled out in documentation: project conception, research plan, technical analysis, Web development (infrastructure), Web design (interface), content development, database design, and so on. Some account of the percentage of effort in the project as a whole should be indicated.

Credit for intellectual contributions /authorship

Project teams have to work collaboratively, and the research activity unfolds within the implementation; it is not separate from it. But the responsibility for the research question and the intellectual contribution of each participant should be made clear in documentation. This should include a description of how the project was shaped by design decisions, discipline-specific knowledge, and technical expertise.

Info architecture/institutional cyber-infrastructure/systems administration

Decisions about information architecture and design are crucial parts of the project. Knowing where the work will sit institutionally, how it will be supported and in what server environments, and how the software and/or platforms for content development will be chosen is at the foundation of the project. It is also necessary to know who will configure the server infrastructure, administer the systems, install the software (and keep it up to date), and back up the content.

Open-source software and technology transfer

Development of tools and platforms is one of the foundation stones of Digital Humanities projects. It is in the interest of the common enterprise of teaching and learning for software to be understood as a community resource with source code shared so as to enable support and development by the user community as a whole. In general, projects should be built with an eye toward fostering common solutions and shared platforms, though there may be times when one-offs serve a specific purpose. How does the project allow for the documentation and transfer of code, tools, platforms, and applications?

Documentation

Documentation of the structure and design of a project is an essential piece of the work. Too often this is ignored. Documentation is essential for continuity of the project after its initial start-up, and it is an important contribution to the field, as well as a way for others to repurpose the design. Development processes should be documented; functional and technical specifications should be documented; system requirements for the project should be documented (for example, which browsers and versions are supported; what plug-ins are required); database entities and relational schemata should be documented; and, finally, code should be documented, including the publicly available code libraries used in the project, licensing agreements or user agreements (especially for APIs), and the intended operations of individual modules, with author attributions.

Audience, user considerations

Making clear who the audience for the project is and how its members are engaged in its development is important, even if the research is driven by an individual scholar's curiosity or agenda. Projects without audiences or users are silos into which work and resources disappear. User-testing is often a critically necessary part of the refinement of the project's interface and navigational features.

Compliance with all legal regulations

Digital Humanities projects must follow Americans with Disabilities Act (ADA) standards in their design and must be compliant with intellectual property and copyright restrictions. The latter are, however, to be applied with a clear understanding of the right to fair use, the not-for-profit character of nearly all humanities research, and the contribution that such research makes to the knowledge and recognition of cultural objects and heritage.

Publishing/dissemination models

Getting attention for a digital project requires putting it into view in an online venue, getting it reviewed, and creating visibility within a scholarly community and among potential users and future contributors. Projects should have a plan for dissemination and publication. Projects built with and from communities have more buy-in than projects built by single scholars. Digital projects should not "rebuild the wheel" but instead strategically assess and, where possible, take advantage of existing software solutions, platforms, or tools. Both the future of humanistic learning and the ability of humanities scholarship to matter in society at large depend upon the unrestricted circulation of scholarly knowledge; accordingly, the least restrictive licenses should be the norm.

Assessment criteria

A project should have its metrics of success and failure stated explicitly. These might range from creating a project that proves a concept or demonstrates a design principle to a project that sets a goal of digitizing and marking up a particular amount of material or engaging a specific community in online discussion and discourse. Having clear goals and milestones is useful as a way to assess the relation between resources and results.

Conversation with multiple fields

Is the project in dialogue with other works in its field, both those traditionally conceived as well as those realized in digital media? Do the authors understand and reference other research and digital projects as models? How does the project situate itself within the intellectual development of a given field or fields?

Sustainability

However experimental its technology base, preservation strategies are a defining feature of good project design. Digital assets are fragile by nature, and this fragility needs to be addressed from the outset by means of a mid- to long-term preservation strategy. What is the plan for sustaining the digital project? Where will it be housed and maintained institutionally? How will those resources be sustained? What will it cost to continue the project, if it is open-ended, and what possible sources of revenue are there for this support? The labor of staff, students, and consultants as well as the costs of hardware, software, and other materials need to be taken into account, not to mention the intellectual commitments of the primary researcher and community of advisors and contributors.

Transparency

All funding sources, whether monetary or in-kind donations, should be disclosed in the various outputs to which a Digital Humanities project gives rise.

SPECIFICATION 3

CORE COMPETENCIES IN PROCESSES AND METHODS

What are the basic skills essential for being able to do Digital Humanities work? How can such projects be supported within an academic or institutional environment? This advisory lists the fundamental elements necessary for the creation of digital research projects. The specific competencies will vary by field and discipline and not all projects require all of these competencies.

All digital projects have technical, administrative, and intellectual aspects to their production. As tools and platforms designed specifically for the Digital Humanities become increasingly available, building custom-designed projects will only be justified if a new tool or platform is part of the development or if the project has some demonstrably unique elements that require a one-off solution.

TECHNICAL

Web development, infrastructure, server environment, interface design; choices about tools, platforms, software, and hardware.

Familiarity with data types and file formats

On what basis are decisions about file formats and data types made?

Database knowledge

If a database is part of the information architecture, what type is it? How will it work and why is it needed? What are the entities in the database, what are their attributes and relationships, and how will the objects be queried and sorted? Is the database open-source, proprietary, and/or licensed? What data sets will be used in the project and who controls them? What kind of permissions and rights will govern the data sets?

XML structured data

What schema or version of XML is being used and why? Is it used for mark-up or just for metadata?

Metadata standards

What process of metadata selection was used and how does the metadata standard suit the project and its disciplinary field as well as its institutional home? Are the metadata standards compliant with existing standards in the field?

Scripting languages

To what extent are scripting languages used in the project and how are they suited to the server and administrative environment in which they work, as well as to the tasks to which they are put?

GIS platforms and spatial data

Tools for spatial mapping and analysis have been developed within geographical disciplines for professional use but other more popular tools for mapping (like Google Earth) have a lower threshold for use. What are the spatial (and temporal) aspects of the data and how will these data be appropriately marked up for analysis? How will they be displayed within a mapping or GIS system, and what are the research questions that can be tested with such systems? Are the data already "spatial" and, if not, is this process automated or does it involve manual geo-rectification of materials (whether maps, historical photographs, videos, or oral histories)? How will this be done, by whom, and with an eye toward what standards for visualization and sharing within and across geo-browser applications?

Virtual simulation tools

Virtual worlds and three-dimensional modeling are tools for creating immersive environments for historical research and presentation. Again, what tools, software, and systems are being used and for what ends? What standards are being followed and how will various communities of practice engage with the models, simulations, and virtual worlds? Into which existing platforms will the models be placed and what kind of constraints do these platforms have?

Existing and emerging platforms for content management and authoring

How will the project manage existing content and support the growth of new content? Who are the authors of this proposed content, and how will they input it? Will they need to be technically savvy or does a browser interface enable their participation? What content management systems are used in the infrastructure or repository? Do the content management systems enable data to be shared across platforms and repositories?

Interface design as knowledge modeling

How is content displayed in the interface and how does a user navigate this content through the interface? What is the interface model and how does it express the knowledge model of the project and support its mission?

Game engines

Game economies have a role to play in scholarly work as well as in entertainment. Understanding the way game engines might be incorporated into a project to support participation is useful in certain circumstances.

Design for mobility and diversity

Does the project have dimensions that will make its content available on mobile applications or allow it to be repurposed for use in multiple contexts? Will the project work on different platforms? Will it work across cultural, linguistic, and social divides? Is the project ADA compliant, or does it have limitations for use by persons with disabilities?

Custom-built vs. off-the-shelf

Is part of the project's research the designing and building of a platform or tool, and if so, can this work be repurposed or generalized from its customized use for a broader audience? If off-the-shelf solutions or standard software systems are being used, how were they chosen? Many times, Digital Humanities projects will be a combination of these two approaches, using existing APIs, standard content management systems, or blogging engines that can be variously customized and extended to address the specific needs of a project.

INTELLECTUAL

While the most visible intellectual element is usually the content, it is important to recognize that Digital Humanities projects present arguments and knowledge experiments in many different ways, often contributing to the creation of new knowledge through complex interactions, visualizations, data and data structures, and even code. Digital Humanities projects are not just about the content (although this is often primary), but also about the design of multiple levels of knowledge and argument from the operations on the back-end database to the front-end access points of a user interface.

Cross-cultural communication

Has consideration been given to the ways in which the design of the project will work cross-culturally? Is it meant to engage communities whose language and/or cultural orientation will be varied?

Generative imagination

Is the project generative and will it continue to create new content, dialogue, debate, and engagement, or is it largely a packaged repository of content meant to be viewed and used but not altered through contributions or extensions? Both of these are worthwhile and serve different needs, audiences, and intellectual goals.

Iterative and lateral thinking

How might the project change over time, and how will reflections on its limitations be used to improve each iteration? Can the project "play well" with other projects by sharing data through Web services frameworks or code modules through code-sharing repositories?

ADMINISTRATIVE

Resource allocation, reporting lines, clear job descriptions, goals, and outlines of responsibility for all involved are crucial and should be spelled out in a memorandum of understanding, at the very least.

Intellectual property

Have rights and copyright clearances for intellectual property been managed and documented? The terms for use of content should be posted clearly on the site and the contact information for inquiring about the use of intellectual property easy to locate.

Institutional circumstances

What is the institutional home for this project and who will be responsible for its maintenance after the project is built? Costs and impacts on human and material resources should be assessed.

Sustainability, funding, and preservation

Long-term plans for sustainability can include migration of the project into an institutional repository, or archiving on a server or paid service provider, or creation of a revenue stream and business model for its ongoing support and maintenance. Collaboration with institutional entities, particularly libraries and data repositories, will be necessary for preserving data created for and by a Digital Humanities project. Can the data be "outputted" easily from the project and archived in standard formats that are widely readable? What kind of data management plan has been created and how will it be implemented? Are there any privacy or security concerns that need to be addressed?

SPECIFICATION 4
LEARNING OUTCOMES FOR THE DIGITAL HUMANITIES

While core assessment standards remain continuous with those of traditional classroom-based humanities pedagogy, the Digital Humanities recognizes the importance of additional outcomes produced by hands-on, experiential, and project-based learning through doing. Digital Humanities pedagogy emphasizes teamwork and implies an increased role for peer assessment, as well as attention to a widened set of skills beyond text-based critical thinking and communication. Outcomes emphasize the ability to think critically with digital methods to formulate projects that have humanities questions at their core. Among the learning outcomes for the Digital Humanities, we prioritize the following:

Ability to integrate digitally driven research goals, methods, and media with discipline-specific inquiry

Acquire and demonstrate new fluencies from working within and navigating across various information platforms to conceptualize and carry out discipline-specific research. In practice, this means bringing together the traditional tools of humanistic thinking (interpretation and critique, historical perspective, comparative cultural and social analysis, contextualization, archival research) with the tools of computational thinking (information design, statistical analysis, geographic information systems, database creation, and computer graphics) to formulate, interpret, and analyze a humanities-based research problem.

Ability to understand, analyze, and use data

Demonstrate ability to synthesize data from multiple sources and harness multi-modal and multimedia technologies to produce digital arguments. Create capacity to formulate a research problem or question that lends itself to a computational approach. Develop ability to analyze problems by applying digital methods to humanities-based data and to interpret the results of digital analysis and computationally produced outcomes in a critically significant way.

Develop critical savvy for assessing sources and data

Judging the reliability of information and knowledge presented in a digital environment requires skills of discernment that examine the source, the authority, and the legitimacy of the digital material. With regard to data, this means examining how they were obtained, marked-up, stored, and variously made accessible to end-users.

Ability to use design critically

Understand the importance of knowledge design in communication, project development, and long-term preservation of digital data in ways that go beyond competence to a critical understanding of tools, their uses and limitations. Develop ability to use computational design thinking to produce forms of argument and expressions of interpretation.

Ability to assess information and information technologies critically

Interrogate digital, visual, and multi-modal information as evidence and critique its formation and validity. Critique the digital features of publications for a) scholarly relevance, b) best practices (e.g., online footnoting and citation, transparency of sources and data), c) attribution, d) authority and argumentative rigor. Understand and critique the epistemologies, worldviews, and structuring assumptions built into digital platforms, technologies, visualizations, and even computational languages.

Ability to work collaboratively

Think across disciplines, media, and methodologies on multi-authored research projects, project proposals, reports, and presentations aimed at both academic and nonacademic communities. Work in teams and participate in peer assessment. Acquire knowledge of the development life cycle of a Digital Humanities project and the ability to understand the needs and priorities of each phase of development. Meet aggressive deadlines and produce completed, fully functional digital prototypes, products, research tools, and publications. Identify and assess specific contributions and roles in collaborative projects for the purposes of peer review and intellectual credit.

SPECIFICATION 5

CREATING ADVOCACY

Among its other activities, digital scholarship asserts the possibility of changed relations between consumers and producers of cultural work. Listed here is a set of considerations for addressing the cultural significance of humanities work, of transforming individuals into prosumers with critical insight into the workings of digital platforms. It also contains a handful of crucial points on which to advocate for Digital Humanities as a field.

Value of the cultural record

Humanistic scholarship is engaged with the production, preservation, and interpretation of the cultural record. Gauging the value of legacy materials and vetting the value of contemporary contributions is essential. In what ways does the project contribute to the cultural record (through preservation of materials, through interactions among contributors, through modes of public engagement, and so forth)?

Humanistic values/cultural significance and legitimacy

Demonstrating the value of interpretive methods and fundamental humanistic values as a counter to those of managed culture is an essential part of advocacy. How are the values and perspectives of the humanities a central part of the contributions of the project? What does the project contribute to the cultural record and how is this record legitimated (and by whom)?

Expanded notions of community and participation

For whom is this project of value and how are they engaged in its production, reception, or preservation? What notions of community and participation are central to the project? How is participation opened up, managed, and facilitated? How are decisions about permissions for participation, inclusion and/or exclusion, made and who makes them? And what are the limits, liabilities, and challenges that remain for participation without restriction?

Ability to analyze modalities of organization and presentation

Skills for understanding the ways media organize and present arguments are the foundations of informed use of information in any environment. The specific characteristics of digital media—in all their multiple, hybrid, and overlapping forms—need their own languages of assessment.

Reflexive awareness of coercive regimes

All media conceal as well as reveal the rules according to which they include certain kinds of expressions and prevent others. What is possible in any given digital space or project and what is not? We must be reflexive, dialectical thinkers aware that any "solution" always prevents certain questions and problems from arising, while privileging the very ones to which it is the answer. All technologies are coercive in some respect, and many have become so naturalized that we no longer consider them coercive but rather self-evident and necessary. It is up to digital humanists to denaturalize these technologies and create fissures for new, imaginative possibilities to come about.

Thinking beyond the ideologies of templates and structured discourse

How do we read the embodiment of power dynamics and relations in the organization of structured spaces and processes? The digital environment structures its ideological expression in the graphical interfaces, the data types, the database relations, as well as in the content of each project. Epistemological defamiliarization—the "making strange"—is an important feature of modern critical thought. The force of delight, surprise, and even alienation in the face of innovative inventions are the enlightening elements of contemporary imaginative thought. What can be shown to wake us from our passive consumption? And how do new ways of knowing, engaging, and designing become the very means to provoke inquiry, generate thought, deepen values, and contribute to the cultural record of our species?

From passive consumer to active prosumer

The role of reader and viewer varies from that of a consumer of material on display to that of a critically informed and discriminating prosumer of cultural materials. How does the project facilitate productive, critical engagement rather than passive consumption?

Creation of citizen-scholars and scholar-citizens

Many projects support the substantive participation of amateurs, scholars without professional affiliation whose expertise in a field is highly developed, informed, and driven by intellectual passion. In what ways does the project integrate (and also evaluate) a multiplicity of perspectives and knowledge-creators? How do scholars—traditionally conceived—become engaged with a broader pubic citizenry, and, similarly, how are citizens engaged in the intellectual project of knowledge creation as scholars?

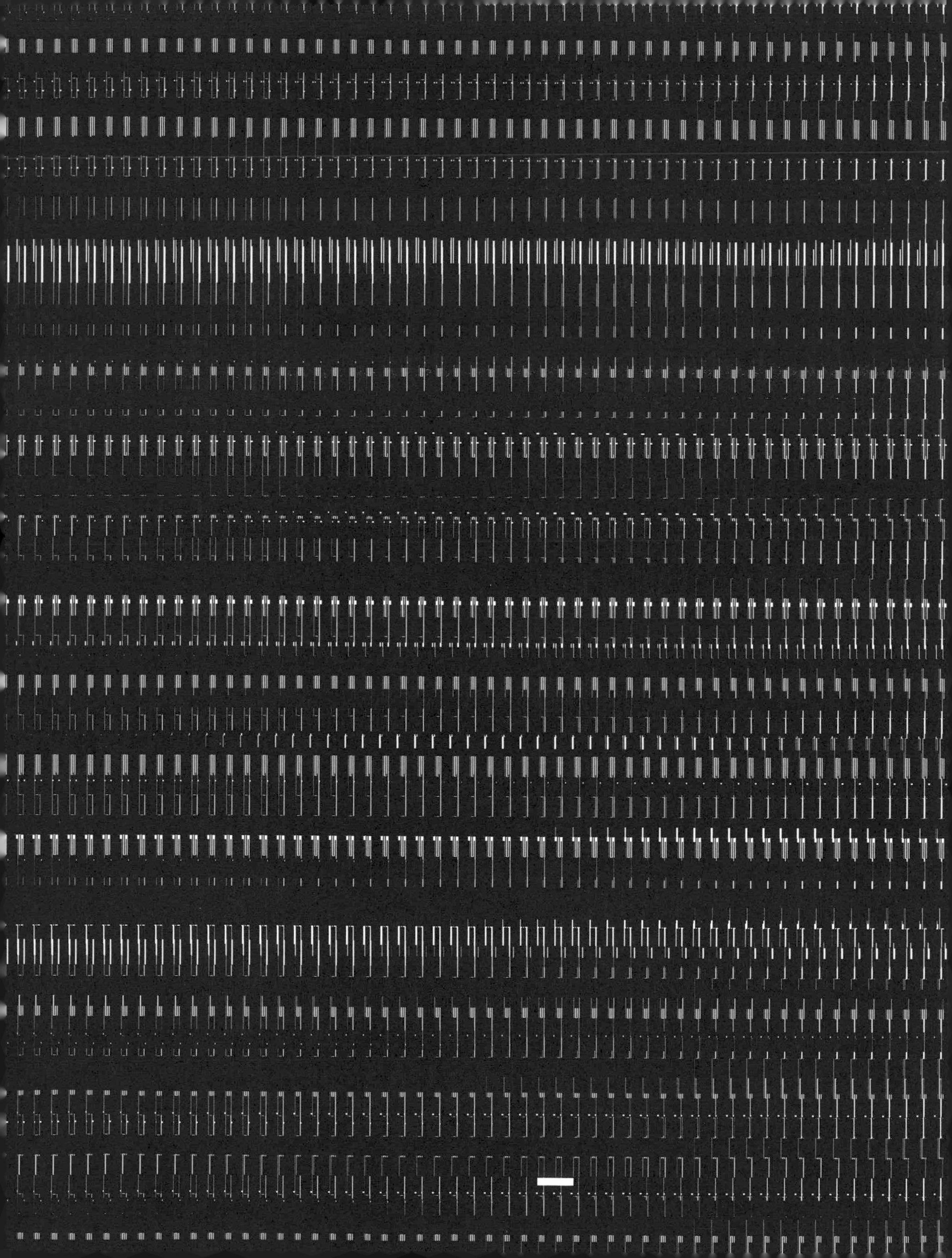

THIS BOOK is a metalogue: a dialogue that assumes the form of that which it discusses. In the present case, we knew that collaboration would be a key theme, so we entered into the construction of this volume with the meta-issues not only in mind, but also in flux. We've noted elsewhere that we did not want this to be a collection of disparate essays, or even a whole with individually signed parts. Instead, we strove to create a consistent, if choral, voice. We also wanted the book's design to be integral to its writing, to acknowledge the inseparability of form and content. From our very first working session we each contributed to the shape of the book, thus we are all listed as full co-authors, a signal that multiple types of knowledge formation require multiple modes of authorship.

The first step toward crafting this voice was, of course, the composition of the team itself. Here the evolution was organic. Each of us had a past working relationship with at least one other member of the team, and we found ourselves coming into ever-increasing contact thanks to a series of allied initiatives concerning the impact of technology on the academy. The convergence accelerated when Jeffrey came down from Stanford to spend the 2008–09 academic year at UCLA co-teaching a mixed reality seminar with Todd. The seminar, entitled "What Is(n't) Digital Humanities?," was funded by the Mellon Foundation and was part of a Mellon-sponsored initiative at UCLA to support transformative approaches to the humanities. During this period they co-wrote the first draft of the "Digital Humanities Manifesto" on Commentpress, opening it up initially to a few other contributors, including both Peter and Johanna, and, subsequently, to contributions, annotations, and even fulminations from seminar participants and the entire World Wide Web.

The following academic year, with the support of the University of California Humanities Research Institute, Peter organized the conference "Nowcasting: Design Theory and Digital Humanities," in which all five co-authors participated and the team convened for the first time. That year, Jeffrey was transitioning from Stanford to Harvard; Todd, Johanna and Peter were steering a new Digital Humanities undergraduate minor and graduate certification through UCLA's academic senate; and Anne was co-writing "Digital Learning, Digital Scholarship and Design Thinking" with Holly Willis. The next year, with the UCLA Program established and metaLAB launching at Harvard, all five of us gathered in Cambridge for "Digital Humanities 2.0: A Conversation About Emerging Paradigms in the Arts and Humanities in the Information Age": an evening in which we each presented our own work within a unified framework. In other words, we had long been testing out our ideas and developing a rapport with one another while at the same time marshaling foundation and institutional support—from the departmental level all the way up through university-wide initiatives.

For two days before the event, we held the first of several charrettes. A few words of explanation are perhaps in order. In the 19TH century, architecture professors at the École des Beaux-Arts in Paris were known for conducting a workshop exercise known as the charrette, in which small groups of students were given a design brief to resolve within a short time frame. As the deadline approached, a charrette (or "cart") made its rounds to collect each group's designs. Students were often observed leaping into the cart along with their submissions, working passionately to add finishing touches, even as the cart bounced along the streets of Paris. We looked to the charrette, an abiding feature of architecture and design training even today, as a fitting means to undertake the writing of a book about the centrality of design to the Digital Humanities. The challenge was to design the book conceptually and graphically in a form that emerged from the ideas.

We worked at the Harvard Graduate School of Design with white boards, laptop computers, and a projector whose long cable we threw back and forth to whoever needed to project something. We began by identifying the areas we wished to frame and then broke into alternating groups to develop headers into lists and lists into structured chapters. These were then posted on white boards as we took turns typing out expanding outlines of the book. Guided in part by Anne's persistent attention to the shape of our arguments and organization of text by theme and format, the design infrastructure began to emerge in the process. By the time we took the dais at the Harvard Humanities Center on the evening of the third day, it felt as if we'd collectively entered the state of focused motivation that social psychologist Mihaly Csikszentmihalyi calls "flow."

To keep the flow flowing we considered a number of production models. We thought of books developed by means of crowd-sourcing, like McKenzie Wark's GAM3R 7H30RY, an online initiative carried out with the Institute for the Future of the Book in 2006 that came out as *Gamer Theory* with Harvard University Press in 2007, and Kathleen Fitzpatrick's *Planned Obsolescence: Publishing, Technology, and the Future of the Academy* which was put through an open peer-review process on MediaCommons before its publication with NYU Press in 2011. Though some of the early ideas for the book appeared in condensed form in the "Digital Humanities Manifesto," we wanted to keep the face-to-face mindmeld alive throughout the entire writing process, experimenting with a variety of private document-sharing platforms and relying upon regular meetings either in person or via Skype.

We imagined a shared-access manuscript as the best analogue to what we had been developing in person, but found that the main platforms were less robust than expected, and, more importantly, that the proliferation of iterations hurt more than the transparency of the compositional process helped. After starting with Google

Docs, we ended up using email and tracking changes in attached word processing documents to create a round-robin writing-and-editing loop. When one person finished drafting a chapter or section, he or she would pass it on to another, who would edit and amend it, and so on. As the round robin proceeded, the book grew with both a speed and unified character that came as a surprise to all of us.

By the time there existed a beta version of the whole, the full manuscript was run through another round of editing by team members. Each iteration was passed on with changes visible to all. The next editor/author/designer in the sequence treated this as a "new" version, looking back at other changes only briefly before moving forward with the iterative writing process. The most uncanny effect of the process was running across lines, paragraphs, or whole sections you yourself had authored but which now were subtly tweaked or appeared in unfamiliar new contexts. We came to analogize this to crossing paths with an amiable ex-lover at a party. There was familiarity and affection, but also a new sense of remove.

While the manuscript was still underway, Anne spent the summer of 2011 at Art Center working on a research project called "MICRO MEGA META." The project investigated the future of scholarly production through the creation of speculative prototypes and design fiction. Both Johanna and Peter spoke to her graduate student researchers about the overarching issues and how they related to the book project. The student researchers, in turn, worked with drafts of the book to design digital environments built upon humanities values. This interaction led to a second charrette at Art Center which was devoted to thinking through the structures of the text and how they might be embodied through the design of the book.

Making the shift from the linear vertical scroll of word processing software to the spatiality and recto-verso of the codex altered the rhythm and organization of the text. We worked with a reader in mind; revisions made the rounds, sections were dropped or altered or moved. Lastly, Anne and Peter spent two days "writing to the design"—editing and embellishing to make the most of the semantic changes brought about through line breaks, recomposed information hierarchies, navigational maneuvers, and spatial relationships. Though Anne created the actual page layouts, the book's design had been underway since that first meeting at Harvard.

The book in your hands is the result of several years of collaborative composition, design, and writing. It will have future lives and iterations as a transmedia artifact, as it migrates into various digital forms and gives rise to its own generative scholarship. We see this book as a beginning, an opening to create and re-create that deep linkage that we call *Digital_Humanities*.

We offer this brief inventory of references and resources as entry points into the expanding field of Digital Humanities. The book's ideas are informed by a vast network of individuals, projects, and organizations that have built the field as it exists today, only a handful of which are cited in the text. Rather than provide a map or a bibliography that represents that network, or any of the networks that make the humanities digital, we refer you here to a list of living resources (which are always subject to change).

In that spirit, individual institutions, labs, centers, and projects are not listed here, given that many online compendia exist and these areas are developing so rapidly that any print work categorizing them risks instant obsolescence.

Research tools, technologies, and platforms

The links below point not only to digital tools and libraries, but also to initiatives concerned with emerging specifications and best practices.

ARTStor Digital Library www.artstor.org

Bamboo DiRT (Digital Research Tools) provides a fairly comprehensive inventory of digital research tools organized by category (ranging from data analysis and text mining to visualization and mapping); it is an integral part of Project Bamboo dirt.projectbamboo.org

Creative Commons creativecommons.org

Europeana www.europeana.eu

Fair Cite Initiative faircite.wordpress.com

Fedora Commons www.fedora-commons.org

HathiTrust Digital Library www.hathitrust.org

Mukurtu www.mukurtu.org

Open Access Directory oad.simmons.edu

Project Gutenberg www.gutenberg.org

Public Knowledge Project pkp.sfu.ca

Spatial Humanities spatial.scholarslab.org

Text Encoding Initiative www.tei-c.org

Voyant Tools voyant-tools.org

Associations and institutions

The following list represents thousands of scholars and centers worldwide, providing leadership and community as well as technical and infrastructural support. Foundations and scholarly societies, such as the MLA, AHA, and ACLS, have also invested in research, teaching, and institutional formations in support of the Digital Humanities.

Alliance of Digital Humanities Organizations www.digitalhumanities.org

Arts-Humanities Net www.arts-humanities.net

Association for Computers and the Humanities www.ach.org

Association for Literary and Linguistic Computing www.allc.org

Council on Library and Information Resources www.clir.org

Digital Library Federation www.diglib.org

Digital Research Infrastructure for the Arts and the Humanities www.dariah.eu

Electronic Literature Organization eliterature.org

HASTAC (Humanities, Arts, Sciences, and Technology Advanced Collaboratory) www.hastac.org

Institute for the Future of the Book www.futureofthebook.org

Institute of Museum and Library Services www.imls.gov

MediaCommons mediacommons.futureofthebook.org

NEH Office of Digital Humanities www.neh.gov/odh

THATCamp: The Humanities and Technology Camp thatcamp.org

Collections, series, journals, and forums

Numerous online bibliographies exist to help scholars address their research and teaching needs, including ones developed and maintained by the institutions previously listed. Edited collections can also provide excellent points of access. Recent publications include:

David Berry, ed., *Understanding Digital Humanities* (New York: Palgrave Macmillan, 2012)

Matthew Gold, ed., *Debates in the Digital Humanities* (Minneapolis: University of Minnesota Press, 2012)

The MacArthur Foundation, Reports on Digital Media and Learning, (Cambridge, MA: MIT Press, 2009–11), available online at: www.scribd.com/collections/2346520/John-D-and-Catherine-T-MacArthur-Foundation-Reports-on-Digital-Media-and-Learning.

Susan Schreibman, John Unsworth, and Ray Siemens, eds., *A Companion to Digital Humanities* (Oxford: Blackwell, 2004), available online at: www.digitalhumanities.org/companion.

Several publication series have been launched by university presses, as well as partnerships between university presses and foundations that envisage the creation of digital publishing platforms. Following are just a few of the growing number of specialized journals, forums and discussion groups dedicated to Digital Humanities work that can be found online:

Association for Computers and the Humanities Q&A digitalhumanities.org/answers

Digital Humanities Now digitalhumanitiesnow.org

Digital Studies/Le champ numérique www.digitalstudies.org

Humanist Discussion Group www.digitalhumanities.org/humanist

Literary and Linguistic Computing llc.oxfordjournals.org

MediaCommonsPress mediacommons.futureofthebook.org/mcpress

Vectors: Journal of Culture and Technology in a Dynamic Vernacular vectors.usc.edu

Sources cited in the text

Aristotle, *The Poetics*, introduced by Francis Fergusson (New York: Hill and Wang, 1961), section IX.

Yochai Benkler, *The Wealth of Networks: How Social Production Transforms Markets and Freedom* (New Haven: Yale University Press, 2006).

John Berger, *Ways of Seeing* [orig. 1972] (New York: Viking Press, 1995).

Isaiah Berlin, "The Divorce between the Sciences and the Humanities," in *The Proper Study of Mankind* [orig. 1974] (New York: Farrar, Straus and Giroux, 1997), 326–58.

James Boyle, *The Public Domain: Enclosing the Commons of the Mind* (New Haven: Yale University Press, 2008), 182.

John Seely Brown and Paul Duguid, *The Social Life of Information* (Cambridge, MA: Harvard Business School Press, 2000).

Bill Coleman, "Thought Leader: Valley Veteran Bill Coleman On Failure And the Guild of Entrepreneurs," interview by Tom Foremski, March 5, 2009, archived at: www.siliconvalleywatcher.com/mt/archives/2009/03/thought_leader_7.php

Kathleen Fitzpatrick, *Planned Obsolescence: Publishing, Technology, and the Future of the Academy* (New York: NYU Press, 2011). Earlier versions available online: www.plannedobsolescence.net

Michel Foucault, "The Discourse on Language," in: *The Archaeology of Knowledge*, trans. A.M. Sheridan Smith (New York: Pantheon Books, 1972), 224.

Jürgen Habermas, *The Structural Transformation of the Public Sphere: An Inquiry into a Category of Bourgeois Society*, trans. Thomas Burger, with Frederick Lawrence (Cambridge, MA: MIT Press, 1991)

N. Katherine Hayles, design by Anne Burdick, *Writing Machines*, (Cambridge, MA: MIT Press, 2002).

Immanuel Kant, "An Answer to the Question: 'What is Enlightenment?'" in: *Kant: Political Writings*, ed. H.S. Reiss (Cambridge: Cambridge University Press, 1970), 54–60.

Friedrich Kittler, *Discourse Networks 1800/1900*, trans. Chris Metteer with Chris Cullens (Stanford: Stanford University Press, 1990).

Lev Manovich, "Cultural Analytics: Analysis and Visualization of Large Cultural Data Sets," CALIT2 White Paper, 2007, archived at: www.manovich.net/cultural_analytics.pdf

Scott McCloud, *Understanding Comics: The Invisible Art* (New York: HarperCollins, 1994).

Marshall McLuhan and Quentin Fiore, *The Medium is the Massage* [orig. 1967] (Berkeley: Gingko Press, 2005).

Franco Moretti, "Conjectures on World Literature," *New Left Review*, 1 (Jan–Feb. 2000): 54–68.

Theodor Nelson, *Computer Lib/Dream Machines* [orig. 1974] (Redmont, WA: Tempus Books, 1987).

Richard Stallman, from a transcription of a 1986 lecture at the Swedish Royal Institute of Technology, 1986, archived at: www.gnu.org/philosophy/stallman-kth.html

Vectors: Journal of Culture and Technology in a Dynamic Vernacular (About): www.vectorsjournal.org/journal/index.php?page=Introduction

Beatrice Warde, "The Crystal Goblet," first delivered in 1930 as "Printing Should be Invisible," in: *The Crystal Goblet: Sixteen Essays on Typography* (London: Sylvan Press, 1955).

Mark Weiser, "Ubiquitous Computing" (1988 at the Computer Science Lab at Xerox PARC). www.ubiq.com/ubicomp

Wikipedia Statistics, accessed online at: en.wikipedia.org/wiki/Special:Statistics

DIGITAL_HUMANITIES